Praise for *Wilted Wings*

"If you think, as I did, that the problem of lead poisoning in birds is mostly solved, think again. In McTee's brilliant, heart-breaking first book, he illustrates through personal accounts of his experience with birds of prey, that lead poisoning is an ongoing problem and taking a terrible toll. I highly recommend *Wilted Wings* to all naturalists and hunters." —Stephen Bodio, author of *An Eternity of Eagles*

"A heartfelt call to action, Wilted Wings makes the compelling case that our choices matter—for humans, our winged fellow hunters, and many other wild species. Hunting lead-free is a win for us all."
—Tovar Cerulli, author of *The Mindful Carnivore: A Vegetarian's Hunt for Sustenance*

"Wilted Wings brings us into the mountains pursuing big game. As a scientist and one of the nation's leaders in the fight for switching to non-lead ammo, McTee studies pathways by which raptors and other scavengers are exposed to lead. In this page-turning memoir, Wilted Wings provides the undeniable solution to lead poisoning in raptors." —Vince Slabe, Ph.D., Wildlife Biologist and Science.org contributing author

"Wilted Wings is the most comprehensive and readable explanation I've found on the science of lead poisoning in raptors and other scavengers, and on why hunters should consider switching to nontoxic bullets, as I did several years ago. Highly recommended for anyone interested in how modern hunters are leading our nation's conservation efforts." —Tom Dickson, Editor, Montana Outdoors, the Magazine of Montana Fish, Wildlife & Parks

WILTED WINGS

A Hunter's Fight for Eagles

Mike McTee

Riverfeet Press
Livingston, MT
Bemidji, MN
Abingdon, VA
www.riverfeetpress.com

WILTED WINGS: A Hunter's Fight for Eagles
Mike McTee
Non-fiction: Nature/Science
Copyright 2022 © by the author
Edited by Daniel J. Rice
All rights reserved.
ISBN-13: 979-8985398816
LCCN: 2022937684

This book is available at a special discount to booksellers, librarians, and educational institutions.
Contact the publisher for orders: riverfeetpress@gmail.com

Cover Design: Jordan Hoffmaster
Cover Photograph: Estelle Shuttleworth
Typesetting and book design by Daniel J. Rice

Some of these stories originally appeared in these publications:
Journal of Mountain Hunting—Lead Bullets: Hunting for Clarity in a Controversy
Montana Naturalist—Spying on Our Connection to Scavengers
Missoulian—Lead poisoning kills Golden Eagle, hunters adapting
Strung—Nomads of the Sky and Earth

PHOTO CREDITS
Estelle Shuttleworth: Chapter 1, Chapter 2, Chapter 3, Chapter 4, Chapter 5, Chapter 6, Chapter 7, Chapter 8, Chapter 10, Chapter 11, Chapter 12, Epilogue
Travis Booms: Chapter 9
Mike McTee: Further Reading about Non-lead Ammunition
Wild Skies Raptor Center: Prologue

Contents

For Bridgette

A Hunter's Fight for Eagles

WILTED WINGS

Mike McTee

January 2021

PROLOGUE

The golden eagle hopped off the logging road on frostbitten knuckles. Its talons—weaponry that can slice open prey ranging in size from prairie dogs to pronghorn—were clenched into a ball. The sluggish wingbeats of the bird couldn't lift it to the lowest branches of the young pines.

Hannah and David had just left Missoula, Montana, to cross-country ski in the Blackfoot Valley. When they pulled off the highway onto the logging road, they spotted the struggling eagle and knew their skis wouldn't touch snow that morning. David parked the car, and Hannah phoned a biologist at Montana

Fish, Wildlife and Parks. The poor connection garbled their conversation, but the biologist told her to catch the eagle if she could. Meanwhile, they'd call the rehabilitators down the road at Wild Skies Raptor Center.

Hannah and David slipped off their jackets thinking they might swaddle the eagle like a baby. As they edged forward, they realized the eagle couldn't summon the force to flee or fight. Hannah placed her hand on the eagle's back as if to introduce herself. With the bird still, David eased his palms over the wings, pressing them tight against the bird's chest. He lifted the eagle, one of fastest beings on earth, now reduced to a sick and stumbling creature. Its head hung as if a dumbbell dangled from its throat. Hannah and David recognized the symptoms and suspected the source. They wondered if lead had laced the veins of another eagle and dragged it to the ground.

***

I talked with Hannah a day later. We'd been friends for years, both worked conservation jobs, and hung around the same bunch of hunters and anglers. Hannah described her eagle encounter in a simple phrase: "It was awful." I carried those words with me for a week like a stabbing headache before calling Brooke Tanner, the head rehabilitator at Wild Skies, asking if I could visit. On the phone, Brooke confirmed the eagle was fighting lead poisoning. Despite me studying the topic for most of the last decade at MPG Ranch in the nearby Bitterroot Valley, I still hadn't seen an eagle clutched in the final throes of lead's sinister grip.

That afternoon, I drove from Missoula along the Blackfoot River toward Wild Skies. I parked beside one of the many sheds placed between the pines in their yard. A bald eagle perched inside a barn that had been retrofitted as a raptor enclosure, or mew, as a falconer would call it. Deer skulls and elk antlers decorated some of the buildings.

I threw on my jacket, slid on a beanie, and walked toward the long, one-story house. Brooke exited the front door, followed by her assistant, Jesse Varnado. Jesse was holding a golden eagle by its legs. The bird's two-inch long talons looked like black grizzly claws.

"This is Rhonda," Jesse said, looking down at the bird as I approached. His glasses failed to hide the worry on his face.

"Oh, that's right," I replied. "I forgot you had her," realizing this wasn't the eagle I came to see. Brooke and Jesse were clearly juggling a variety of patients.

Four years ago, my collaborators at Raptor View Research Institute had caught this adult female, fitted her with a GPS transmitter for a study, and named her Rhonda. After release, Rhonda hunted the Bitterroot Valley's foothills, rising to the higher elevations as the snow receded. By summer, Rhonda was perching on whitebark pines growing from the highest granite summits in the Selway-Bitterroot Wilderness. Her transmitter quit sending locations in late September.

Nearly a year later, the unit rebooted and sent a stack of points from a ridge in the wilderness. A biologist recovered the transmitter but found no signs of Rhonda. Three weeks ago, Raptor View recaptured her at a trapping site, knowing it was Rhonda based on her numbered leg band. She was emaciated. Her lead levels maxed out the LeadCareII instrument. Something had broken two of her toes. That same foot was infected. The biologists drove her to Brooke.

"That swelling is terrible," I said, staring at a toe that was inflamed to three times its normal size.

"It's actually looking a lot better," Brooke said. "Do you see this scarring?" She ran a finger along a sore, about the width of a thin wire that bridged the toes. "It could be from a snare trap."

Jesse walked Rhonda to a raptor enclosure to put her away while I hung back with Brooke.

"How's she doing?" I asked.

"Her lead levels are down, but I'm worried about the bone infection," Brooke said.

Infection often accompanies trapping injuries. With an eagle's foot tight in the trap's wire or jaws, each jump and twist cuts a deeper slice into the bird's scaly feet. Those same feet might then land on a carcass brimming with bacteria. "I got a golden eagle once with trapping trauma that died within twenty-four hours by infection," Brooke said. Others had to be euthanized because the injury was so severe, or because it's illegal to keep or release a raptor with only one foot. Brooke has seen raptors with shattered and fractured legs, broken feet, no feet, localized infection, systematic infection, broken wings from cars, broken wings from bullets, dehydration, emaciation, anemia, and lead poisoning. The tattoo on her right arm sports a collage of raptors, blending a red-tailed hawk, northern saw-whet owl, a pair of peregrine falcons, and of course, a golden eagle. Under her compassionate eyes runs an undercurrent of toughness that a rehabber of fifteen years must own.

Brooke toured me around her property, explaining that most of the sheds were raptor enclosures. Like the bald eagle in the barn, all the inhabitants experienced some form of traumatic injury that rendered them unfit to fly or find food. We approached one enclosure, and a Swainson's hawk side-stepped toward me along a wooden pole, probing my face like a dog expecting a treat.

"In the summer, I'll feed him grasshoppers through the wire," Jesse said, catching up with us. "When he sees me come out with the insect net, he gets excited." Swainson's hawks winter in Argentina. On this particular hawk's first autumn migration, a utility pole zapped him with electricity fifty miles east of Missoula, grounding him for life. With only one good wing, he couldn't even flutter between perches. The Swainson's enjoyed being around people, though, making him an ideal bird to show at educational events. Brooke and Jesse hoped that by exposing people to raptors, others would appreciate birds of prey enough to protect them—or at least not shoot them when they glide over a chicken coop.

"Should we give the eagle fluids?" Jesse asked, speaking now of the golden eagle Hannah and David had found.

"Yeah," Brooke replied. We walked back to the house and went inside. In a small room to the right, they kept American kestrels and saw-whet owls. To the left was their laboratory, outfitted with workbenches, metallic trays, syringes, and sophisticated-looking tabletop instruments. At the end of the lab, a closed door separated us from their Intensive Care Unit.

"Do you want to see her before I grab her?" Jesse asked.

I nodded. The walls of the ICU were piled with animal crates except for a gap in front of the window. Drab blankets covered the cages. In the middle of the room sat an infant crib with blankets draped over the rails and bottom. The golden eagle lay with her chest flush to the crib floor. Brooke said the eagle was a female, based on its large footpad and heavier weight. Most raptors are sexually dimorphic, with females outsizing males. Dark brown plumage dressed the eagle's body, meaning she was an adult anywhere between five to thirty years old. In the crib, the eagle's wings drooped, and her head hovered inches above the bed. The nictitating membrane of her eyes swept across the brown orbs. The eagle opened then shut her beak. And again. Open. Closed. Open. Closed. I turned to Jesse.

"Is it normal how she's opening her beak like that?" I asked.

"It is when they're in bad shape," he replied.

Jesse hunched over the crib and hoisted the eagle. He embraced her almost as a father holds his newborn. Jesse's hands grasped its legs, giving it support from below. His other hand held the neck, and his fingers stroked the feathers under her beak. The eagle's head still sagged, not limp, but low. She stared out the window at a carpet of snow and the puzzle-shaped bark of ponderosa pines. The regal animal whose shadow could send jackrabbits dashing to cover had withered like a flower after a September frost. The outside light brightened her eyes. I gazed at the mottled landscapes of her brown irises. Ever since seeing a golden eagle's eyes up close nearly a decade before, I always wondered what type of mind swirled behind those deep, black pupils.

"Brooke, what do you see when you look into a golden's eyes?"

"Intelligence," she said. "Like their wheels are turning. Like they're processing things. They're unlike other raptors. The majority of goldens seem to realize you're helping, and they begin cooperating."

Jesse walked the bird toward a chest freezer by the front door. Brooke tried to slide a falconry hood over the eagle's head. "Too small," she said, grabbing another hood from a makeshift hat rack on the wall with a dozen more options. "You've got a big head," she said as she fitted the eagle with a larger hood. This one sported faux rattlesnake skin.

Jesse laid the bird's back on the chest freezer. The eagle's tail feathers looked odd, almost like a beaver's tail. Jesse and Brooke had pressed her tail feathers between x-ray film and taped them together. It helped prevent her feathers from being destroyed while lying down. Feces covered the x-ray film and caked the feathers around her vent.

"Normally, we give the birds a quick bath, but with this one," Brooke stopped, "we just want her to survive, then we'll give her a bath. There's no reason to exhaust her more than we need to."

"At what point do you decide to euthanize an eagle?" I asked.

"You can tell when they've given up," Brooke said. "They won't acknowledge you. There's no fight in them. But this eagle still makes eye contact. It's swallowing, too. That's a good sign."

Eagles pick up food and throw it back to eat. This eagle couldn't swing that exaggerated movement, so Brooke and Jesse had been hand-feeding her quail and venison. Wild Skies had a permit to salvage roadkill for raptor food. They collected thirty deer carcasses the previous year. Although Jesse called the permit a "lifesaver," it didn't lower their annual bill for raptor food. More and more birds come in each year.

With the eagle's back on the freezer, and Jesse holding her feet, Brooke opened a wing. She leaned in with a syringe, parted feathers, and began drawing a blood sample. Seven days had passed since they last tested this bird's lead levels and red blood cell count.

"The vein is collapsing," Jesse whispered. That was a sign of dehydration, even though the eagle received fluids twice daily.

Brooke finished taking the sample and walked into their lab space. I heard a centrifuge rev up, and she turned on her tabletop lead analyzer. Brooke returned with a large syringe that had flexible tubing running to a needle. Jesse lifted the eagle from the freezer and turned her face down. Brooke separated the feathers behind its shoulders, first with the brown outer plumage and then the wispy white down underneath. Once she opened a clear path to a sliver of skin, she inserted the needle.

"This is an electrolyte solution," Brooke whispered. The eagle remained motionless as nearly two fluid ounces poured in.

"Raptors absorb subcutaneous fluids," Jesse said. "It causes less stress to inject it under their skin than to stick a tube down their throats."

"When an eagle comes in with lead poisoning and we give it fluids, it usually perks up. It's kind of a false hope." Brooke swapped her massive syringe for a tiny one.

"This is the chelator," Jesse said. Brooke pushed the plunger. This particular chelator was calcium EDTA. It's a large molecule that binds to metals, like lead. The bird excretes the resulting compound. Otherwise, the lead would continue to flow through the eagle and be stored in bodily tissues, ravaging the bird's nervous system. Brooke and Jesse administered chelating treatment daily.

"We like to inject it along with the fluids," Brooke said, now walking into the lab and setting the syringes on a metal tray. "The other way to do it is to inject it into their muscles. But I think it causes too much pain. Can you imagine your muscles getting poked ten times in one week with needles?"

Jesse moved the eagle to the crib and closed the ICU door behind him. Brooke approached a workbench to see what the blood tests revealed. "Her red blood cell count is up. That's good." Brooke showed me a slender glass tube, no bigger than a toothpick, where she mixed the blood with a serum to measure packed cell volume. "Do you see the color in that serum? That yellow indicates liver damage." Brooke sighed. "That's disheartening."

Brooke stepped over to her LeadCareII instrument, dispensed a drop of blood on blotter paper, and inserted it into the sample port. The instrument beeped. The LeadCareII was designed to rapidly measure lead levels in children, but rehabbers and biologists used it for wildlife. The instrument beeped again and Brooke read the result. The eagle's lead level had dropped by 90%. Still, the bird had almost no control of her body, and such a high dose of lead likely caused irreparable damage. It's like putting out a house fire just before the building collapses—the structure is still standing, but a breath of wind might topple it to ashes.

According to Brooke, other rehabbers in western Montana had

taken in eagles that month. Becky Kean at Bozeman's Montana Raptor Conservation Center worried that lead exposure caused permanent damage in the golden eagle she had just received. It breathed with an open mouth, indicating a possible respiratory issue. A rehabber in Kalispell had just watched lead poisoning kill a bald eagle.

"The birds we get in with lead poisoning are usually picked up along roads," Jesse said. "How many die in forests, where no one finds them?"

Jesse's remark reminded me of a golden eagle they recovered the previous month. The only reason they tracked her down was because the eagle carried a transmitter. The previous year, that bird had collided with a car, breaking her pelvis near Philipsburg, a former mining town between Missoula and Butte. Wild Skies rehabilitated her for six weeks. Before her release, my collaborator, Rob Domenech at Raptor View, outfitted the bird with a GPS transmitter.

Brooke and Jesse had access to the transmitter data, so they watched their former patient cruise the river valleys around Missoula. In the spring, she flew north over British Columbia and into Alaska, casting a tiny shadow over massive glaciers that flowed for miles below stony peaks. The eagle soared past the Brooks Range and stopped where the Arctic Ocean lapped against North America's edge. When the first real darkness fell in September, the eagle sailed south on air currents above a tapestry of deciduous shrubs and tundra. She passed Mount Cleveland, the tallest peak in Glacier National Park, before taking her winter residence in western Montana. Around the winter solstice, the eagle ended her 6,000 mile journey in a forest, dying from lead poisoning within twenty miles of where Brooke and Jesse helped heal her pelvis.

"Brooke," I asked, "what do you feel when you work to save these birds but so many die?"

"It's sad. It's hard." Brooke paused, as if the memory of each poisoned eagle was dragging her toward grief. "Some people think that birds shouldn't be rehabbed. But just recently, Rob Domenech sent us a photo of a tagged eagle that we rehabbed four years ago. It feels good to know some of them are still out there. It keeps me going. If we're not working with these birds, who's going to tell their story?"

From inside the ICU, the golden eagle emitted a chirp as delicate as smoke rising from a snuffed candle.

"I hate hearing that," Jesse said.

"At least it means she's still alive," Brooke replied.

Some mornings Brooke and Jesse enter the ICU to see their patient perched on the side of the crib. "That's when we know they're ready for the next step," Jesse said.

When they entered the ICU the morning after my visit, everything we'd seen in the depths of the eagle's eyes had flickered and fallen dark.

Nine years earlier...

1. DELAYED IGNITION

On the edge of the massive Bob Marshall Wilderness Complex in Montana, I sat at the bottom of an avalanche chute, aiming my crosshairs at a bull elk with a gun that wouldn't fire. When I squeezed the trigger, a single, sharp click vibrated through my hands. The elk, with five points on each antler, stood 100 yards uphill and seemed unbothered by the noise. He kept plucking grass from wet soil. A steady rain had fallen overnight, and gray fog enshrouded the upper half of the mountain. Water wicked into my pants, but I couldn't move to sit on my boot. The beargrass growing above hardly hid me from view. The bull stepped forward, now only four paces from disappearing in the forest.

With my rifle resting on a bipod, I worked the action and ejected the ammunition into my hand. I held the silver cartridge between my fingers like a cigarette and studied the bottom. The primer, or the piece that ignites the gunpowder, had a shallow dimple. The rifle would only shoot if the firing pin in the bolt struck the primer like a snappy punch instead of a feeble jab. I tucked the round into my jacket and chambered the next one. My crosshairs found the knot behind the elk's front shoulder. I flared my nostrils to suck a breath. As the exhale poured out, my finger tightened on the trigger. *Click.*

The bull kept feeding as his antlers swayed with each bite. Maybe he mistook the click for a woodpecker tapping a tree. I cycled the bolt, squeezed, and registered another failure. The bull raised his head, sensing something was amiss. His eyes inspected between the clumps of grass surrounding me, but he never anchored his stare to my face. Still, I had snuck into his bedroom just to clank around. If he realized I wasn't a woodpecker, he'd turn to the trees. He'd find a thick, dark forest where brittle branches were strewn across the dirt. Any misstep in my stalk would snap those twigs and trigger his security system. My rifle needed to revitalize itself in a hurry, otherwise, we'd be packing out frustration instead of meat.

I glanced toward my dad, who stood fifty yards below at the tail end of the avalanche chute, behind a thicket of spruce. That's where we had spotted the bull before I crept over. Beside him was my friend Beau and his wife, Julie. My dad and Julie both furrowed their brows. Beau lifted his hands and shrugged as if asking why his ears weren't ringing from a gunshot.

My eyes returned to the elk, and I thought about the dimpled primers. The gun had laid under the tent fly all night. Maybe the cold temperature had thickened the oil in the bolt and made it sticky. I rested the butt of the rifle stock on my leg and began dry firing, cycling the action each time. If I built heat within the bolt, maybe the firing pin would free up. At the sixth click, the bull settled a piercing stare on my troubled eyes. I needed to chamber a cartridge, but the elk wouldn't tolerate a flailing arm or ammunition fumbling between my fingers. I inched my hand to my jacket pocket for a fresh round. The fleece muffled the rustling as I pressed the ammunition into the magazine. I slowly raised the weapon over three breath cycles and aimed. The trigger shivered with another click.

With a few swift strides, the bull left me sitting in an empty avalanche chute, holding a cold, lifeless gun. I sighed into the fog, rose from the soggy soil, and plodded toward my audience. Beau and Julie accepted me with sympathetic eyes. My dad stared at the gun like he wanted to break it over his knee. "You should throw that gun off a cliff!"

"The damn firing pin must be gummed up," I said, dropping the rifle on my backpack like it was a piece of trash. "I suppose this is the problem with bringing only one gun."

"How old is that thing?" My dad asked.

"At least 30 years." I turned to Beau and Julie, who didn't know the whole story. "I've been borrowing this rifle from our friend. It's worked the last couple seasons. It worked this week at the range." Beau and Julie dug their hands deeper into their pockets, knowing their words wouldn't alleviate my throbbing disappointment.

The four of us stood in silence, exchanging glances, each contemplating the unspoken question: Do we keep hunting with a gun that won't fire?

I crouched and rummaged through my backpack, filling my pockets with fresh rounds. They clinked against the dimpled ammunition. "Since we're here, we might as well keep trying." I released the bolt from the rifle and zipped it inside my jacket. "If the bolt warms up, maybe it'll work. Who knows, the bull might pop out in the next avalanche chute." I shouldered my backpack and rocked it side to side until the straps settled. "I'm going up there."

Julie brightened. "We'll keep a lookout down here." Beau and my dad nodded.

A stubby subalpine fir grew mid-way up the second avalanche chute, offering the only cover. I climbed up and leaned my pack against its narrow trunk. I sat on my boot this time, extended the rifle's bipod, and rested the stock on my thigh. The adrenaline that had previously warmed my nerves began to subside. My wool pants felt like damp towels resting on my legs, and my elbows trembled in the cold. I turned downhill for a distraction. Beau and Julie were milling around. My dad was combing the mountain with his binoculars, hoping to catch an animal sneaking behind a cloak of spruce.

As the chill crawled over my skin, despair began stalking through my thoughts. If I had understood guns better or maybe cleaned the bolt, we'd be skinning my first elk. Rifle hunting was still somewhat new to me. I had hunted since before I could grow armpit hair, but my dad and I were traditional archers. Our recurve bows and wood arrows handicapped us to shots within thirty yards, if the opportunity arose at all. We rarely took home venison, and I learned that traditional archers hunt with patience and tenacity rather than bloodlust or greed. When I relocated from western Washington to Missoula, Montana, to eventually pursue a degree in chemistry, I moved in with a native Montanan who kept venison on the table year-round. Securing a reliable supply of free-range backstraps and burgers attracted me, so when archery season ended, I traded the bow for a rifle and my chest freezer began to brim with meat.

But on that avalanche chute, as I watched the fog creep up the mountain, I realized that rifle hunting required a degree of attention I'd largely ignored. Shooting a gun with a powerless pin was like shooting a recurve bow with a loose string. I knew nothing about the bullets I had bought, either, other than they were lead and would pack a hefty blow to an elk. I looked downhill again.

My dad was missing. Julie spotted my face, and then she and Beau signaled me down with sweeping waves of their arms. I sprung up, grabbed my gear, and began hurdling over beargrass.

"I think your dad sees the bull," Beau whispered.

My dad was standing on the other side of the chute with a curtain of trees at his back. When he saw me, he stretched his arms above his head like he was holding a beach ball. I had known that sign since elementary school: antlers. He pointed at me and traced a route with his finger along the base of the chute. Countless avalanches and tree throws had scoured the surface of the lower chute to steep bedrock. That slope break would conceal my traverse. I stepped lightly. Grinding stones or initiating a rockfall would sabotage the hunt before I could retest my rifle. Once across, I skirted the forest uphill to my dad. He was standing below a chest-high rock ledge, smirking. I peeked over the ledge and saw why. Upslope from where the first bull had been grazing stood another bull—a bigger bull—with six points on each antler.

I drew the bolt from my jacket pocket, hoping it had warmed, slid it into place, and chambered a cartridge. The rock ledge made a firm rest for my bipod. I leaned into my stock to absorb the recoil I wanted to feel so badly. The elk crowded my view through the riflescope as the crosshairs settled. *Click.*

The elk remained at ease, so I snatched a handful of rounds from my pocket and rolled them in my palm. They all had dimpled primers except for one. I slammed the fresh cartridge in the gun and trained its aim on the elk's shoulder. When I touched the trigger, a small missile of hot lead and copper carved through the mountain air at nearly 3,000 feet per second. The elk felt the jolt, spun, and fled toward the timber. He collapsed before the smell of gunpowder had cleared the air.

My dad grabbed me and rattled the dumbfounded stare off my face. "Those elk were on a conveyor belt!" he laughed.

***

This landscape was both treasured and feared for its grizzly bears, so the four of us quickly gathered our gear and hiked up to the 600-pound animal to begin butchering. We unzipped knives from backpacks. My dad pocketed his ceramic sharpener. We'd need to hone our blades after dulling them on the bull's thick hide. I laid white canvas bags on the ground nearby for the meat.

As the shooter and by unspoken agreement, I made the first cuts. Beau held up a back leg, and I pierced a shallow entry on the elk's belly, just long enough for me to dig two fingers underneath the skin. I pressed against the warm stomach to create space. My knife sliced toward the sternum, and the skin peeled open like a jacket being unzipped. We were now staring down at the stomach, liver, intestines, and the spiderweb of caul fat that wraps them.

I pushed the organs aside with my free hand and slashed past the diaphragm to the chest cavity. The lungs felt sopping wet with hot blood. With a series of jabs, I severed the esophagus and windpipe. A few cuts later, we yanked over 100 pounds of soft organs from the bull's splayed belly and rolled the steaming gut pile downhill.

With the elk's insides now outside, we started skinning the bull and removing meat. My blade ran between the spine and the loin, freeing a backstrap as thick as a python. I held up the back leg while Beau worked his knife around the ball joint. We hauled the leg over to a tree and hung it with rope, where my dad carved muscle from the femur. Meanwhile, Julie freed up small chunks of meat Beau and I had missed. "Here's some good burger!" she said, dropping pieces into the bag. We had turned a nearly botched hunt into a smooth-running disassembly line.

Beau began freeing the front leg by running his knife through the armpit muscles. "Oh cool, here's the bullet," he said, twisting it free from the underside of the bloody leg. Beau dropped it in my palm. "That's a cool souvenir," he said. The bullet had taken a new form. Instead of being a miniature missile of lead, it looked like someone had pressed a melted candle against a wall. I zipped the bullet away in my backpack.

After three hours, we'd filled five bags with boneless meat. Now, we needed to schlep the protein and antlers down the mountain to our base camp. We'd spend the night there, and in the morning, load the game cart for the final walk out. Before that, Beau and I would shuttle most of the meat in two trips between the kill site and the camp. We started down the mountain ahead of Julie and my dad, who were running just one trip.

The packing went fast, and on my second trip down with Beau, we left behind what wasn't destined for our dinner plates. The rib-cage glistened red next to the internal organs and bones. I didn't consider which scavengers would find the remains other than grizzlies, who roam my mind whenever I hunt near them. Gray jays could have been perched on nearby spruce trees, patiently waiting for supper. Once they fluttered down and gleaned scraps, their harsh chatter could draw in larger predators. This time of year, thousands of golden eagles were migrating over these mountains, searching for prey on windswept ridges, watching for scavengers skittering around bloody skeletons, hunting for an easy meal to fuel their southbound flight.

2. CAPTURE

The blue Toyota 4Runner rumbled up an icy road at MPG Ranch in western Montana. It crested the ridge and stopped. Rob Domenech, the Executive Director of the Missoula-based Raptor View Research Institute, cranked down the driver's side window. He reached to the back seat for his spotting scope and mounted it to the window, squinting his left eye and peering toward the swath of snowy grassland that lay below. A quarter-mile out, a golden eagle perched atop a snag sticking out from a draw.

"You got an age on that bird?" Adam Shreading asked from the passenger seat.

"Looks like a sub-adult," Rob said in his gravelly voice. "Fingers crossed it comes down to feed."

A bloody landing pad lay below the eagle. Rob and Adam had placed a road-killed deer there the day before, staking it to the frozen ground. Dead animals make good eagle bait, but carcasses don't last long. Coyotes had tested the strength of the stakes in the morning and dragged bloody organs across the snow. Magpies and ravens were attempting to peck the bait to bones, but knew their place in the hierarchy of birds. The instant the golden eagle jumped from the snag and spread its wings, the corvids scattered in a frenzy of black and white.

"Here we go," Rob said as the eagle touched down. The bird tucked in its wings and swiveled its head to take in potential threats. Standing on a dead white-tailed deer made the bird vulnerable. Coyotes could charge. Other eagles could swoop. This eagle overlooked the camouflaged box sitting ten feet away.

The eagle's beak tore bits of flesh from the deer's back leg with violent jerks. It shifted its feet one at a time, slowly beginning to face away from the camouflaged box. Adam leaned forward with anticipation, watching through his binoculars. Rob, who was now holding a remote control with a telescoping antenna, pressed its single green button. With a loud crack, three weighted foam projectiles launched from the box, casting a twenty-foot by twenty-foot net over the eagle. Startled, the eagle leaped skyward but hit the woven trap.

Rob jammed the 4Runner's stick into first gear and stomped the accelerator, testing his tire's grip on the slick road. The engine howled across the quiet grassland before skidding to a halt 100 yards from the entangled eagle. Both doors flung open, and Rob and Adam sprinted through ankle-deep snow to the bird.

The webbing had wrapped around the eagle's feathers and legs. It could hardly flutter a wing, but its deep brown eyes tracked the biologists as they neared. The eagle's talons were open, ready to dig deep into a soft forearm. A bald eagle had once sunk a talon into Rob's finger right to the bone, sending him to the emergency room for powerful antibiotics. Here, Rob grabbed the eagle's legs with a bare hand to control its feet. With his free hand, Rob helped Adam remove the net. Once the webbing cleared the eagle's beak, Rob slipped a leather falconry hood over its head and cinched it tight using his teeth. The darkness calmed the golden eagle, and Rob and Adam hiked it back to the 4Runner.

Two miles away at the Orchard House, I was typing on my laptop at the kitchen table, writing about pollution at the ranch's abandoned shooting range. A car flashed past the window. A moment later, the screen door creaked open and slammed shut. Adam entered the kitchen, his winter boots pounding against the hardwood floor. Rob followed, pressing a golden eagle against his thick, olive sweatshirt.

"Hey Mike," Rob said. "Do you mind giving us a hand with this bird? We're a little understaffed today."

"Yeah, of course." I rose from the kitchen table.

"Hold its legs like two drumsticks," Rob said. "Keep the talons up."

My hands replaced Rob's, and I brought the golden eagle to my torso, amazed that a bird with a six-foot wingspan could be as light as an infant. Its feet were clenched, but the curved daggers on each toe still frightened me. Small patches of white speckled the eagle's brown chest and its tailfeathers. Rob and Adam aged it at three years old based on that plumage. They unlatched a plastic container and removed calipers, rulers, and other instruments. Rob opened the jaws of the calipers and measured the length of the eagle's beak. He separated the talons and measured the length of the footpads. He read off measurements for Adam to write on a clipboard, but otherwise, he worked quietly, hoping to minimize stress on the bird. Rob opened a small plastic container with rows of metallic leg bands. He grabbed one that had the number 0709-01880 etched on its side, slid the band on the eagle's leg, and riveted it closed. Rob then wrapped the bird's feet with a Velcro strap, carefully slipped the hook of a fish scale through the strap, and lifted the 9.7 pound bird. Rob guessed I was holding a female based on the bird's size. DNA analysis later confirmed the sex.

Adam grabbed the bird's legs and laid its back on the hardwood floor in front of the refrigerator. He spread the right wing. Rob took a syringe from the plastic container, lowered himself to his knees, and leaned in. He parted feathers with an alcohol swab to expose a vein, inserted the syringe, and drew up the plunger. He removed the needle and dispensed the blood sample into a plastic tube. Adam pressed a cotton ball against the vein to stop the bleeding.

With Adam still holding the bird, Rob rummaged through his banding kit and found four wing tags. They looked like ear tags for cattle, but their vinyl construction made them lightweight and pliable. Each tag had the number 168 painted in white. Rob settled on the floor beside the eagle. He held a tag on either side of the bird's left wing, just above its wrist, and reached for an instrument that resembled pliers. "I'm going to punch these tags between a thin layer of skin at the patagium," Rob said. He steadied the instrument on the tags. "We're the only ones who use blue wing tags with white numbers, so if anyone sees an eagle with one, we banded it." He squeezed and the tags thumped into place.

Once Rob tagged the second wing, he walked the bird outside into the cold mountain air. Adam opened the bird's beak and fed it small nuggets of venison. "This was like an alien abduction," Rob said. The eagle's neck grew as her crop filled with meat. "We want to send her off with a good meal."

Rob removed the falconry hood, and the previously docile eagle turned ferocious. Its feet snapped open and closed with terrifying speed. Just before Rob pushed the eagle into the cloudy sky, the bird's fierce brown eyes seared into mine. Her irises held a topography as intricate as the terrain she hunted. But it was her penetrating black pupils that would forever linger in my mind.

***

After Rob released the golden eagle, the mystery of what those brown eyes would see absorbed into my imagination. If the bird flew west, it would cruise past bald eagles perched in cottonwoods along Montana's Bitterroot River. It would scan the floodplains between thick ponderosa pines that still showed the scars from where Salish Indians long ago harvested the sugar-dense inner bark. Further west, the eagle might watch cars drive Highway 93 to and from Missoula, following a centuries-old path that the Lewis and Clark Expedition called a road even back in 1805. Past the noisy thoroughfare, the eagle could ride thermals lifting from the Bitterroot Mountain's foothills until it gazed down on the granite summits that stand 5,000 feet above the valley floor.

If the bird remained on the east side of the valley, cruising along the rolling grasslands that rise toward the forested Sapphire Mountains, it would encounter some of the fastest growing suburban sprawl in Montana. The eagle would soar past custom houses built on desolate ridges where irrigated lawns have replaced ground squirrel colonies. Eventually, the eagle might take perch on a ranch separating subdivisions, hoping to feast on bloody afterbirth from the calving that stains the February snow.

The biggest mystery was where the golden eagle would drift come spring when many raptors migrate. Rob's tagged eagles had been spotted as far south as Chihuahua, Mexico, and as far north as the Arctic Circle in Alaska. We never learned the migratory habits of the eagle I held, although Rob and Adam spotted her the next year. She was alive and well, once again feeding on their bait at MPG Ranch.

That eagle changed the way I looked at the sky. My hunting buddies became used to waiting while I paused to identify a raptor cutting across the horizon. I hung a *Sibley's Raptors of North America* poster in the house. I studied its illustrations at night while I brushed my teeth. I enrolled in an ornithology class at the University of Montana for entertainment. On a road trip from Montana to Oregon, my wife, Bridgette, and I stopped for a quick break at The Peregrine Fund Headquarters in Boise. We toured their facility, gawking at a Eurasian eagle owl's humungous eyes and catching every tour and raptor showing they offered. When our growling stomachs signaled we missed lunch, we realized the raptors had seized us for four hours.

And so often, my mind returns to the eagle's stare. Raptors have some of the biggest eyes in the avian kingdom. Eagles have eyes nearly the size of a human's, but they're crammed into an apple-sized head. Those massive optics, along with talons and a beak designed to shred flesh from field mice to dead moose, define raptors. Above their eyes runs the supraorbital ridge, a browline that protects their eyes and gives them the "I could kick your ass but I am busy" look, as the author Brian Doyle put it. And I'm convinced they could. They certainly kick their prey's asses.

On snowy grasslands, golden eagles occasionally ride pronghorns like horse jockeys, stabbing their beaks into the warm muscle as blood dribbles down the animal's side. They chase coyotes from carcasses like a bouncer at a bar. (Except the bouncer doesn't have knives for fingernails). Mating pairs trick jackrabbits as one bird distracts the prey with an aerial dive and the other intercepts it for the kill. They stoop from the heavens at up to 200 miles per hour so they can smack a goose out of formation. Golden eagles use humans to their advantage, too. Near Pikes Peak, Colorado, each time biologists hazed bighorn sheep with a helicopter toward a capture location, golden eagles emerged from the mountainside to attack the fleeing lambs. The biologists gave up after eagles had killed two sheep.

Then there are bald eagles. Technically, they're fishing eagles, but they'd be better classified as scrappy foragers. Once while I fished the Missouri River, an osprey plunged into the water after a spectacular dive and rose with a trout. A bald eagle appeared from hiding and orbited the osprey from behind and below, driving the weighted raptor up and up. At the zenith of their spiral, the osprey released its catch and the eagle dove like a dart to snatch the falling fish. That type of thievery has earned the bald eagle a sometimes unfavorable reputation. "He is a bird of bad moral Character. He does not get his Living honestly," said Benjamin Franklin, who preferred the wild turkey for the nation's symbol. Maybe Franklin never saw a bald eagle hunt.

One spring in Idaho's backcountry, Bridgette and I spotted a snow goose floating down the river. The giant white bird, off its migratory course and alone, caught the attention of a bald eagle. The eagle sprung from its treetop perch and circled over the goose, investigating it for weakness. The eagle saw a chance and swooped. Once its talons were about to bathe white feathers in blood, the goose dunked under water, and the eagle missed. The goose bobbed up and flapped its wings. But the goose didn't rise into flight, so the birds continued this deadly dance over a dozen times. A third of a mile down the river, just before the exhausted goose drifted from our view, the eagle plunged, and its talons grabbed goose flesh. Carrying momentum from the dive, the eagle flew the goose to the rocky shore and clenched its talons back and forth like a cat kneading its paws.

On my journey toward becoming an eagle enthusiast, I finally learned how to distinguish juvenile balds from goldens. Bald eagles don't develop their iconic white head and tailfeathers until they reach four or five years. Until then, they sport shades of brown and charcoal, making them near doppelgängers to golden eagles. Sorting them out is like distinguishing between black bears and grizzly bears—relying on color paints an incomplete picture. I learned to first focus on the facial profile. Bald eagles have enormous bills and bulky heads compared to goldens. Juvenile balds also lack the golden hackles that adorn the napes of golden eagles. While both species can grow white feathers, goldens almost only grow them before adulthood as a patch on each wing and along the base of their tails. In comparison, white feathers can mottle a juvenile bald's body like a middle-aged man's salt and pepper hair.

To help calibrate my raptor identification, I kept tabs on Rob's crew when I saw their 4Runner racing after a raptor. Whenever they had a bird in the hand, I wandered over. Releasing an American kestrel from my palm always beat-out digging for soil samples. Watching Rob and Adam pluck an osprey chick from its riverside nest overpowered my need to tinker with the irrigation pump. Our restoration planting would have to wait for me to gawk at the nestling's goldfish-colored eyes.

I became hooked on raptors, and I wanted more.

***

Before meeting Rob, I had envisioned most birders as retirees dressed in khaki who wore goofy hats. But raptor biologists fit a different stereotype, and at a conference, I heard Rob admit it. "Raptor trappers have a reputation for being rough and gruff—and drunk," Rob said. "Like ice fisherman, but for birds." The drunk portion doesn't match Rob, but he might appear rough and gruff by appearance. He often covers his bald head with a flat top hat, is sturdy like a wrestler, and has practiced martial arts all his adult life. Maybe the pain from decades of leg kicks and chokeholds prepared him for the sting of talons piercing into his hands.

When someone asks Rob how he started Raptor View Research

Institute, he might say something like, "It takes more balls than brains." He grew up in New Jersey with train tracks and factories for neighbors. He could tell which town he was driving through by its stench. Being surrounded by blue-collar workers and steady factory jobs, most kids aspired to operate forklifts as adults, but Rob's path strayed toward the natural world. One of his earliest raptor experiences came by way of his neighbors, who Rob describes as Jersey Hillbillies. They kept a snapping turtle in a pool and a diamondback rattlesnake in their house. Rob even watched their pet raccoon climb a tree to raid a crow's nest. One summer, his neighbors kept an American kestrel in a cage outside the house. Rob caught carpenter ants every day and held them between the small openings of the cage. The falcon flew from its perch to snatch each one. In high school, Rob drew scenes on his desks of golden eagles chasing jackrabbits.

As a teenager, Rob drove west with his family into the Appalachian Mountains to visit the Hawk Mountain Sanctuary in Pennsylvania. Hawk Mountain has an internationally renowned research and education facility that hosts over 60,000 visitors per year. Each fall, visitors can witness thousands of raptors migrating over the forested mountains. On the day Rob visited, raptor biologists emerged from their trapping blind near the observation deck with a Cooper's hawk. They had already taken measurements and banded the bird, so they searched for an impressionable visitor to set it free. Rob caught their eye, and they handed him the crow-sized raptor. Like my experience holding the golden eagle, releasing that hawk proved a formative moment for Rob. Over the coming autumns, he ventured to the summits of Appalachia to watch raptors soar south, eventually befriending a professional hawk watcher who taught him the fundamentals of raptor identification.

After high school, Rob made a migration of his own to Missoula to roof houses with his uncle. Rob kept a pair of binoculars nearby while stapling shingles, and he spotted his first migrating raptor in Montana from atop a house. On his lunch break, he raced down to the Clark Fork River and watched Cooper's and sharp-shinned hawks pass over the cottonwoods. Afterward, he called Hawk Watch International and asked where people were counting hawks near Missoula. Nowhere, it turned out. They suggested he find a site.

Rob began steering his '84 Subaru wagon up rutted Forest Service roads to viewpoints. In 1999, he drove 7,000 miles and scouted thirteen potential sites, all within a two-hour drive of Missoula. One of the sites was on the Rocky Mountain Front, on the southeast corner of the Bob Marshall Wilderness Complex. There, the short-grass prairie of the Great Plains slams against the cliffs and peaks of the Rockies. That abrupt elevation change creates consistent updrafts where winds deflect off ridges, helping raptors gain lift. On one ridge, pockets of timber were interspersed with meadows, which provided unobstructed views to the north. On Rob's first day, golden eagles filled the sky. They were high, low, east, and west. Rob kept returning. On some days, he counted 200 eagles as they billowed from the horizon.

He'd discovered a bottleneck along the migratory highway for golden eagles that were departing Alaska and northern Canada. Those northern latitudes give the birds secluded bluffs and cliffs to nest. The adults hunt plentiful arctic ground squirrels and scoop up unsuspecting willow ptarmigans. Come fall, nearly all of those golden eagles migrate for warmer climates. Eagles have broad wings and tails that equip them to soar with ease. When sunlight warms the land, the land warms the air, creating bubbles of heated air that rise like the goo in a lava lamp. Eagles spiral up those thermals and glide to the next. Some birds settle down around southern Canada. Others fly as far as northern Mexico, potentially cruising past a martial arts aficionado from New Jersey who admires them with pure joy.

The next fall, Rob returned to the Rocky Mountain Front. By his second full season, he had assembled a crew of biologists and began attempting to catch and band eagles. People told Rob that migratory eagles wouldn't feed. But Rob and his team tried anyway, always honing their approach. They dressed a rock pigeon in a talon-proof leather vest and tethered the bird to a rope. Everyone waited nearby in two rectangular, camouflaged blinds hidden among trees. They positioned the blinds 50-yards apart so they could watch the sky from different angles. One blind was equipped to catch eagles. The biologists in the second blind were spotters. Squinting through their binoculars, they picked out golden eagles over ridges a few miles to the north. They communicated with walkie-talkies in a dialect that incorporated nicknames for landscape features. I visited once, and a typical conversation went something like this:

"Hey Adam. We've got an eagle above Caribou." Rob says. "It's about two minutes out."

"Got it," Adam replies from the capture blind.

"It's on a good line," Rob says.

Crouched in the blind to conceal his movement from the eagle, Adam tugs a rope, so the pigeon jumps and flutters. The motion can catch a golden eagle's eye from a mile away, but this eagle is already much closer.

"Thirty seconds out," Rob says. "Get ready."

That's when the eagle might tuck its approach below the ridgeline. Expert hunters rarely arrive in plain sight. The pigeon, a former resident of downtown Missoula's rooftops, flaps around as one of North America's largest aerial predators slices between the trees in a dull roar. In a flash of brown and gold, eight black daggers plunge at the pigeon. Adam pulls a cord, and a bow net releases over the raptor, entangling it as the pigeon dances to safety.

But it took years to develop that technique. The first season, Rob and his crew trapped and banded hawks without issue, but they only caught four eagles in over two months of trying. When eagles did attack, the bird's power and size sometimes destroyed the trapping sets. Older birds occasionally dropped sticks on the setup, a trick Rob says they use to flush prey. Rob figured the birds knew something wasn't quite right. But by making small modifications over the years, Raptor View upped its numbers to about thirty golden eagles caught annually over just six weeks.

A couple of seasons in, and before Rob began trapping at MPG Ranch, his friend and fellow raptor biologist, Bryan Bedrosian, visited the site. Lead poisoning in California condors was becoming a hot issue, supposedly because they were eating spent bullets left behind in carcasses and gut piles. Since Raptor View was catching dozens of golden eagles each fall, Bryan suggested that Rob check the eagles for lead. Nobody had done it before with migrating golden eagles.

Rob and his crew started drawing blood from each bird. After seven years of sampling 178 golden eagles, they found that 58% of the birds had lead concentrations above background—that's the threshold indicating an eagle was exposed to an unnaturally high amount of lead, which is 0.1 parts per million. For comparison, that would be like taking a fleck of metallic lead the size of a grain of salt, shaving off a third, and dissolving that tiny shaving into a one-liter bottle of blood. That almost imperceptible concentration doubles the threshold the Centers for Disease Control and Prevention consider elevated for humans. Worse, roughly 10% of the golden eagles had lead levels high enough to be clinically lead-poisoned. These levels can induce vomiting, anemia, and blindness in eagles. For a visual predator, blindness equals starvation. Seven eagles had concentrations high enough to cause death.

The results shocked Rob, considering the eagles were migrating from mostly untouched tundra and mountain ranges. Plus, his trapping season often concluded in mid-October, days before the start of Montana's five-week general rifle hunting season. If he tested eagles after the hunting season, he wondered, would they be even more "leaded?"

When I weighed the bullet Beau found in my elk, more than 30% was missing.

3. BEFORE UNLEADED GAS

The ancient Greeks and Romans knew lead was good stuff. They mined and smelted so much of it that the troposphere filled with toxic fallout. Greenland ice cores dating back to that era contain four times more lead than what would be natural. The word "plumbing" comes from the Latin word *plumbum*, which means lead. (Hence, the chemical symbol "Pb"). The Romans indeed used lead for plumbing, and they also used it to line clay pipes, aqueducts, and reservoirs.

But lead's versatility went beyond conveying water for the Romans. Is your aging face beginning to wrinkle? Smear lead on your skin! Eye bothering you? Apply a dab of lead-infused ointment!

Does your copper pot give your food a nasty flavor? Line the pot with lead, and your next dinner will taste like dessert! Would you like to sweeten your bitter wine? Just boil grape juice in your lead pot, and *voilá*—you have lead acetate! It's a *killer* sweetener.

Once lead enters the human bloodstream, the body expels some of the toxic atoms through urine and bile. The rest catches a ride on red blood cells and tours the interior, entering soft tissues. As lead marinates in the body and brain, it substitutes for calcium and causes biological processes to go haywire. Lead interferes with energy metabolism, compromises motor skills, and impairs cognitive function. Lead exposure has been linked to a reduction in memory, I.Q., and learning. Anxiety and depression can manifest. Childhood exposure has been correlated with increased rates of violent crime. Scientists once thought that the cognitive problems lessened once the exposure stopped. Newer studies suggest otherwise. Once the damage is done, it seems to be permanent. In high enough doses, lead impairs almost all organs, induces tremors, triggers hallucinations, and sometimes kills.

Lead sneaks into bones and teeth, where it's stored long-term. As a person's bones exchange nutrients over their life, previously deposited lead can escape to the bloodstream. Bone turnover often ramps up when a body is starved of calcium, such as during pregnancy, breastfeeding, and menopause. Lead can even cross the placenta in pregnant women into a developing fetus. Children are most at risk partly because their developing bodies absorb lead faster than adults.

The scientific literature overflows with other disturbing findings about lead. One particularly disturbing study said modern Americans have 1,000 times more lead in their bones than Southwest American Indians did a millennia ago. That study came out in 1991 after cars had burned leaded gasoline for decades. Back in the 1920s, a laboratory at General Motors discovered they could boost the octane of gas by adding tetraethyl lead. The substance also improved engine compression and reduced knocking. Cars then pumped lead from motorways onto gardens, playgrounds, and into the air that steeped inside people's lungs.

Lead-based paint also entered the slurry of environmental pollutants. The lead in paint improved the paint's durability, drying speed, and gave it a long-lasting fresh appearance. Unfortunately, when children plopped paint chips into their mouths, they were rewarded with a taste as sweet as candy, making them want more.

By the late 1970s, after decades of lead bombarding Americans, a staggering 85% of white children and 98% of black children had elevated blood lead levels. (Racial inequality frequently accompanies environmental pollution, severely affecting those who have poor living conditions and nutrition.) Health agencies once set their elevated lead threshold at 10 micrograms per deciliter. That's the same threshold Rob Domenech sets for eagles. But the Centers for Disease Control has since sliced that threshold by more than half for children while also saying no safe level has been identified.

After communities had been drenched in lead exposure for decades, governments began phasing out its use. By 1975, drivers could purchase unleaded fuel widely. By 1996, leaded gas was banned for new on-road vehicles in the U.S. That phase-out had a profound effect on people's health. Between 1976 and 1991, the mean level of lead in Americans' blood dropped by about 75%. It's reasonable to assume this phase-out increased the average I.Q. of Americans, not because it improved their smarts, but because fewer lead atoms were migrating through their noggins. The U.S. government banned lead-based paint in 1978. However, it still clings to walls in older homes, so whenever someone buys or rents a house built before 1978, they must sign a lead disclosure agreement, acknowledging the building may carry a toxic risk.

During the 1990s, kids even learned about lead before they could tie their shoes. In 1996, Sesame Street released their "Lead

Away!" segment. Elmo, the muppet, marched behind kids and an adult who held signs like picketing union workers. "Stay away from peeling paint," they shouted down the sidewalk. "Leave your shoes at the door. Lead is yucky. Wash your hands before you eat."

Oscar the Grouch then sprouts his green head from a trash can. "What is going on here?" he asks. Muppets spend the next 13 minutes singing their mantra no fewer than 10,000 times until the words are stitched to every viewer's brain.

Today, many Americans enjoy a relatively lead-free existence compared to a half century ago. Yet, lead still enters human bloodstreams through old paint, soil, water, and even foreign candies and medicines. The water crisis in Flint, Michigan, illustrates how people can still unknowingly become victims of lead exposure. In 2014, the city changed its water source without adding chemicals that inhibited lead from corroding off pipes. Lead levels spiked in tap water, exposing roughly 100,000 residents to lead, causing President Barack Obama to declare a federal state of emergency and prompting a proposed legal settlement of $641 million.

Even though I was raised on Sesame Street and studied chemistry in college, lead still infiltrated my life. As a teenager standing on the shoreline of the local lake, I crimped lead split-shot to fishing line using my teeth. In college, I tied flies by wrapping lead wire around the hooks for weight. I rowed a drift boat down the Bitterroot River as a lead anchor swung off the stern. And of course, lead spewed from my gun barrels.

Once, I was eating mule deer steaks with my college roommate, Logan, when I bit into something jagged and hard, about the size of a sunflower seed. I spat it out and a piece of metal clanked onto my plate.

"What's that?" I asked, prodding it with my fork. "A bullet fragment?"

Logan leaned in for a look. "Yep," he chuckled.

4. LEAD SHOT, DEAD DUCKS, and POISONED EAGLES to the RESCUE

Despite lead being banned from gasoline, paint, and other applications, it still enters the diets of wildlife. While earning my degree in environmental chemistry at the University of Montana, I worked in a microbiology lab, studying heavy metal exposure in the Clark Fork River. The stream's headwaters flow from Butte, about ninety miles east of Missoula. The surrounding rock once held some of the richest copper deposits and mines in the world. Over a century ago, flooding washed mine tailings and heavy metals into the river, choking out much of the aquatic life. The trouble with heavy metals, such as lead, is they don't degrade. Instead, they're swept further downriver, accumulating in the stream's inhabitants.

My research was guided by Dr. Philip Ramsey, a microbiologist, who also worked as an ecological consultant. Before teaming up with Philip, I thought river microbes were just something that gave people diarrhea. But insects graze microbes from river cobble the way cows graze grass. Those nutrients then climb the rungs on the trophic ladder on up to trout and bald eagles. Sharing a lab bench and microscope with Philip for a couple of years gave me a firm footing with the scientific method and a working knowledge of heavy metals, which would soon come in handy.

In 2009, when I was one semester shy of earning my degree, a landowner that Philip consulted for bought the property that would become MPG Ranch. He gave Philip a budget to establish a lab in Missoula and hire a team of researchers to restore and study the land. Philip left his position at the university, and once I earned my diploma, I moved to the ranch to live and work full-time.

One of our first tasks at MPG Ranch was to deal with a surplus of lead pellets strewn across a floodplain. Previous owners had operated Bitterroot Sporting Clays. The club was similar to trap and skeet shooting, where marksmen take aim at flying disks called clay pigeons, except participants at this property walked a course—sort of like golfing but with a shotgun. The 23-station range snaked under cottonwoods beside the Bitterroot River, contoured around ponds and a wetland, and climbed a grassy hill overlooking the floodplain. Bald eagles nested in a tall cottonwood snag near the first station.

As opposed to rifles that shoot a single bullet at a time, shotguns typically unleash over a hundred pellets at once. The pellets, or shot, often range in size from peppercorns down to BBs, and they fan out from the gun's muzzle to create a cloud of projectiles. Fast-moving game such as grouse, rabbits, and ducks stand tough odds against a skilled shotgunner at close range. For over a century, hunters favored lead shotgun loads because they carried momentum and pummeled prey. After a single bird hunt, hundreds, if not thousands of lead pellets could litter the landscape like hailstones after a July thunderstorm. That's how millions of waterfowl got into trouble with lead.

Birds don't chew their food. Digestion initiates when a meal descends and hits the enzymes and acids in the first chamber of their two-part stomach. Then, the food slides to the second chamber, called the gizzard, which is particularly muscular in waterfowl. Imagine dropping a few sunflower seeds into a coin purse filled with sand and pebbles. Rub the purse between your hands like you were warming your palms on a cold morning. The grit will eventually pulverize the sunflower seeds into digestible particles. That's basically how a duck's gizzard works. Ducks obtain that grit by swallowing sand and stones. Sometimes they gobble up lead shot, too. A single lead pellet can kill a mallard. Six pellets always prove fatal.

By the time the current landowner bought MPG Ranch, a decade's-worth of lead pellets were buried between bits of gravel. They were also submerged in the ponds' muck and cobble. With ducks dabbling over toxic shooting range sediments, we decided to remediate the area. I started by studying its various forms of pollution, earning a Master's Degree in Geosciences along the way. Meanwhile, excavators relocated the loose debris that ducks were most likely to swallow. We transported the fill to an old gravel quarry on the ranch and capped it with topsoil and plants. Now mallards, goldeneyes, and other species kick around a clean pond.

***

It always surprised me that Bitterroot Sporting Clays obtained permits to shoot in a floodplain. The history of waterfowl being poisoned by lead shot goes back over a century. If you've visited Montana's Glacier National Park, you've possibly seen a sizable (albeit shrinking) hunk of ice named after George Bird Grinnell. Grinnell was a pioneering conservationist in the late 1800 and early 1900s, and is highly regarded among hunters for helping Theodore Roosevelt found the Boone and Crockett Club, an organization committed to wildlife conservation. In 1894, Grinnell wrote in the magazine *Forest and Stream*, the publication that would later become *Field and Stream*, that lead pellets from shotguns poisoned waterfowl.

In the decades following Grinnell's article, researchers continued to document widespread lead poisoning of waterfowl in habitats much like the former shooting range at MPG Ranch. A decade after the last bombs had fallen in World War II, lead poisoning was taking an estimated 2-3% of North America's waterfowl every autumn. Ammunition manufacturers had already been developing nontoxic options since before the war, but progress stalled out because duck numbers were actually on the upswing.

The waterfowl rebound had been decades in the making and peaked in the 1950s. If you rewind to early 1900s, many animal populations were in tatters after years of market hunting, indiscriminate killing, and the draining of wetlands for agriculture. The Duck Stamp and Pittman-Robertson Acts of the 1930s are credited for reversing much of that damage, and they still generate millions of dollars today. Waterfowl hunters sixteen years and older are required to buy the Duck Stamp, and 98% of those revenues are directed toward protecting wetland habitat. Meanwhile, the Pittman-Robertson program places a 10-11% excise tax on hunting equipment and firearms. That money gets divvied up among the states and funneled toward programs that benefit wildlife and the people who enjoy the outdoors. As more and more people took up hunting post-World War II, revenues from the Duck Stamp and the Pittman-Robertson began piling up, helping duck and other wildlife populations bounce back, making it easy to ignore a need to develop nontoxic shot.

Although burgeoning hunter numbers bankrolled many wildlife restoration efforts, more hunters meant more lead pellets began lying along waterways. Eventually, all that lead caught up with the waterfowl, and by the early 1960s, the freshly restored populations began declining.

This drop in numbers reinvigorated efforts to produce nontoxic shot. Manufacturers coated lead with different metals. Manufacturers produced solid tin pellets, but their low density meant they wouldn't swat birds as hard as lead. They tried depleted uranium, but the environmental consequences of scattering a slightly radioactive metal across waterways remained unclear. Steel shot finally won out, but it had downsides. Being harder than lead, steel threatened to erode the soft bores inside older shotguns. Its hardness made it a menace to molars, too. A bite into a duck breast hiding a steel pellet could crack a tooth.

The hunting community raised other concerns. They worried that the effective range of steel was inferior to lead and might wound more waterfowl. There were also issues with cost and availability. Not only was steel more expensive, but it wasn't available for small shotgun gauges, which could be potentially discriminating based on age and gender. If nontoxic requirements caused hunters to give up the sport, fewer conservation dollars would be generated through Duck Stamp sales and the Pittman-Robertson program. Opponents to nontoxic shot even questioned whether lead poisoning killed enough waterfowl to warrant a ban. But wildlife managers weren't willing to allow widespread waterfowl die-offs.

To ease hunters into the idea of shooting steel, seven national wildlife refuges began requiring nontoxic shot in 1972. The Fish and Wildlife Service used the opportunity to gauge hunter reactions and evaluate the performance of the new product. As part of the effort, an observer supplied unmarked shells to hunters that either contained lead or steel shot. This program and others like it revealed no evidence that steel wounded waterfowl more than lead.

I bought my first box of steel shot in college before hunting ducks with my roommate, Logan. "The low clouds mean the ducks should fly low," he told me as we parked beside the Bitterroot River as gray light sifted between cottonwood branches. We unzipped our shotguns from their cases and headed downstream, skirting around a man hiding in the willows, overlooking decoys soaking in the shallows. We waded a small channel that detoured from the main current. Water lapped the bellies of our chest waders as we approached a spring that belched from the floodplain.

Logan and I hunched below the riverside grass as we shuffled toward the side-channel. We nodded to one another when a subtle quack betrayed the mallards' location. I stepped toward the slough's outlet and two ducks slapped their wings into flight. My muzzle tracked the rear bird and Logan's gun swung toward the front. With two successive bangs, both mallards splashed back into the water, and I concluded steel shot worked just fine.

But three decades before that duck hunt on the Bitterroot River, some stakeholders still questioned steel. In 1976, the Department of the Interior, which operates the Fish and Wildlife Service, released an Environmental Impact Statement that called for phaseouts of lead shot in problem areas across the country. Swift litigation tangled up progress. The National Rifle Association fought against the program, arguing steel shot damaged firearms, wounded more waterfowl, and could cause dental problems. The court shot down the NRA's challenge and their subsequent appeal.

Litigation was a mere hurdle compared to the nearly impenetrable roadblock created by the 1978 Steven's Amendment. The amendment prohibited the Fish and Wildlife Service from implementing or enforcing their nontoxic shot program without the approval of individual states. The only states that grabbed the baton were Wisconsin and Kansas, who passed legislation requiring waterfowl hunters to use nontoxic shot. Illinois, Maryland, and Wyoming did the opposite by banning or restricting *nontoxic* ammunition.

The Steven's Amendment held up progress for over half a decade until bald eagles began dropping dead from lead poisoning.

***

Back when the Mayflower set anchor off New England's shore, the bald eagle population numbered between a quarter to a half-million nationwide. Once the tentacles of European settlement stretched across the continent, bald eagles started facing a barrage of threats. They were shot for their feathers. They were shot to protect livestock. They were shot probably just because they perched too close to someone holding a gun. Waterways and wetlands were degraded and drained, stripping away habitat and prey. When ranchers and trappers laced carcasses with poison to kill wolves and coyotes, bald eagles also writhed in the dirt until their agonizing death. The crumbling bald eagle population elicited the Bald Eagle Protection Act of 1940. The Act prohibited people from killing, selling, or basically disturbing bald eagles in any way, whether they were dead or alive.

Then came DDT. The miracle insecticide helped rid communities of disease-carrying mosquitoes and some agricultural pests. Unfortunately, it also accumulated in insects and ascended the trophic chain on up to raptors. In a bird's body, DDT transformed to DDE, a substance that interferes with calcium making its way to eggshells. Bald eagles, among other avian species, began laying thin-shelled eggs that were so fragile they sometimes cracked under their mother's weight. In 1962, biologist and author Rachel

Carson published her book *Silent Spring*, exposing the shattering damage artificial chemicals can inflict on the environment. The following year, the number of breeding pairs of bald eagles had collapsed to fewer than 500 in the lower 48 states, earning them Endangered Species protection in 1967.

While a team of scientists were testing for organic contaminants in dead bald eagles, they noticed one bird from Maryland had lead shot in her stomach and esophagus. After publishing that finding, more biologists started checking eagles for lead exposure, and they kept finding lead lurking in digestive tracts. Bald eagles were probably snatching waterfowl that swallowed lead pellets or had pellets embedded in their bodies. But to be certain lead shot was killing bald eagles, biologists for the Fish and Wildlife Service selected five eagles that were "unsuitable for rehabilitation and release or captive breeding, but were otherwise healthy and in good condition," and fed them lead pellets. Four of the five birds died. The fifth bald eagle went blind and had to be euthanized. Lead pellets were indeed killing our national bird.

With the aid of the Bald Eagle Protection Act, Endangered Species Act, and the Migratory Bird Treaty Act, the bald eagle unlocked the path toward phasing out lead shot for waterfowl hunting. These statutes provided the Fish and Wildlife Service authority to advance far-reaching measures to reduce bald eagle mortalities. In 1985, they chose regions within eight states where they wanted to implement nontoxic regulations. When five of those states declined to participate, the Fish and Wildlife Service flexed their muscle by offering the states an ultimatum: either you require waterfowl hunters to shoot nontoxic shot or we're closing your 1985 waterfowl season.

The percentage of waterfowl harvested with nontoxic shot in the 1985 season tripled.

Legislation alone wasn't going to change hunter behavior. Starting back in the 1960s, wildlife managers realized outreach programs needed to precede nontoxic requirements. Before those outreach programs became prevalent, however, opinions about steel shot had already proliferated. Hunters bickered about steel's performance. Advocates from each side of the debate wrote magazine and newspaper articles with contrasting information that confused hunters. It didn't help that personnel from conservation agencies, who often interacted with hunters, didn't always know the facts regarding lead poisoning, or they expressed their personal bias against steel.

To curb these counterproductive conversations, the U.S. Cooperative Lead Poisoning Control Information Program was formed in 1982. They ran shooting clinics where hunters could see first-hand the performance of steel. The Fish and Wildlife Service also created a three-person team that were armed with in-depth knowledge about ballistics, waterfowl management, and lead poisoning. They held public forums that were sometimes verbally hijacked by opponents trying to prevent an open discussion. At times, people threatened their safety. But the nontoxic freight train kept chugging.

In the late 1980s, the National Wildlife Federation accelerated the transition by pressuring the Department of Interior to prohibit lead shot for waterfowl hunting across the Lower 48. In response, the Interior then released a phase-out plan, and in 1991, conservationists popped the corks on champagne bottles. Ninety-seven years after George Bird Grinnell wrote about lead poisoning in *Forest and Stream* magazine, 100% of waterfowl hunters in the United States were required to use nontoxic shot. The lead-shot ban saved an estimated 1.4 million ducks in 1997 alone, marking a grand triumph in the history of conservation. Ammunition manufacturers have continued improving their nontoxic shot to fit more shotgun gauges and have produced pellets made from bismuth and tungsten.

Meanwhile, the numbers of bald eagles have flourished enough that their endangered species protections were lifted in 2007. Banning lead shot for waterfowl hunting helped, but so did the banning of DDT, improving their habitats, and other restorative efforts. The Fish and Wildlife Service estimated in 2020 that over 300,000 bald eagles were fishing and foraging in the lower 48 states. Yet, lead poisoning never stopped killing eagles.

5. BRING ME a CARCASS,
HOLD the POISON

A golden eagle barrels from the sky and touches down. It side-steps around the carrion and begins stabbing its sharp beak into a liver the size of a catcher's mitt. Without needing to chew, the eagle gobbles half the purple flesh in four minutes. Its crop (the pouch in its neck that acts as temporary storage) swells as if its throat had grown a goiter. The eagle climbs atop the carcass like a king on his throne. A couple of magpies land and grovel for scraps.

Seeing the eagle become comfortable bolsters a coyote's confidence. The song dog scampers through knee-high grass on quiet paws, but the bird whips around, flares its golden nape, and spreads its wings. The eagle is outweighed but not outmatched. The coyote backs off and begins pacing within lunging distance. But hunger overwhelms the coyote's patience and the bold canine flashes its glistening fangs. The eagle launches at the intruder, thrashing its wings in pursuit. They brawl outside of the game camera's field of view, and when the camera triggers one minute later, the eagle is perched atop the bloody throne.

***

The idea of watching dead animals with game cameras came from a desire to see more eagles. For 14 years, Raptor View had been fitting golden eagles with blue wing tags – like the ones I watched Rob attach to the eagle in the ranch house. Biologists have traditionally marked eagles with metal leg bands, each one etched with a unique number. Those numerics are almost impossible to read even when staring at a perched bird through a fancy spotting scope. Only about 10% of those eagles are reencountered, with most of those instances occurring when someone finds a dead eagle.

The highly visible blue tags, however, have yielded a 23% reencounter rate, with nearly all recorded observations being of live birds. These reencounters help researchers learn about the birds' migration habits, longevity, and overall life stories. Biologists and citizen scientists have spotted Rob's tagged eagles up and down the Rockies and beyond. One golden that Rob tagged on the Rocky Mountain Front ended up in Cody, Wyoming, four years later. Eight years after that, Dr. Travis Booms, a biologist for the Alaska Department of Fish and Game, spotted that same eagle near Denali National Park, tearing meat off a moose carcass.

An eagle's prey options dwindle in the winter. Ground squirrels and marmots, a dietary staple for many golden eagles, are tucked away in underground burrows. Frozen rivers and lakes make for tough fishing for bald eagles. These lean months force eagles to scavenge energy-rich food so they don't waste precious calories chasing and killing prey. Rob began brainstorming with MPG Ranch's head avian scientist, Kate Stone, about how they could reencounter more tagged birds. They knew carcasses and gut piles were meat eater magnets, so they decided to place game cameras near carcasses; they would just need to access more land than MPG Ranch.

Kate saw this as an opportunity to engage different demographics of people and show them the importance of private land to wildlife conservation. Recreationists often praise public land for being outstanding wildlife habitat without acknowledging the value of private land. About 22% of the contiguous U.S. is either leased or owned for wildlife-related recreation, which includes hunting. Kate began networking with landowners and locals. She printed t-shirts with their "Bitterroot Valley Winter Eagle Project" logo. That first season, she secured roughly thirty locations for bait stations up and down the valley, the majority being on private land.

On most winter days, a biologist from Raptor View or MPG Ranch scoured the local highways for roadkill. "I'm a scavenger for scavengers," said MPG Ranch biologist, Eric "Kerr" Rasmussen, as we were driving southbound on the Eastside Highway, with a carcass salvage permit sitting between us and deer legs jutting from the truck bed. We were in route to the day's first game camera, although we also hoped to sniff out a carcass hidden in a ditch. Most often, though, Kerr and the rest of the crew retrieved carcasses already collected by the Montana Department of Transportation, with white-tailed deer comprising most of the catch. (Moving carcasses away from motorways lessens the odds of a scavenger being smacked by a car.)

"I picked up a deer last week that was just smashed, like it was hit by a Dodge Ram Heavy Duty. When I lifted it to the tailgate, intestines started spilling out." With gastrointestinal matter caked on Kerr's pants and shirt sleeves, he sometimes had trouble eating lunch. "I only eat what I don't have to touch." Sardines had become a staple. Not only could he use a fork, the fishy smell helped mask the deer funk swirling the truck cab.

The ditches were free of roadkill, so Kerr and I turned off the highway at the Teller Wildlife Refuge, driving over packed snow toward the river. The truck cornered a clump of ponderosa pines and a half dozen bald eagles blasted skyward from the bait station. Kerr parked and checked the carcass, ensuring the masonry stakes and steel wire were solid in the ground; otherwise, mountain lions and coyotes might run off with the bait, leaving a lonely sight for the game camera. Kerr opened the cover of a camera strapped to a wooden fencepost. He swapped the memory card so the crew and the refuge staff could see the catch.

Studying scavengers might seem like lowly research. After all, scavengers do dig their faces in maggot-filled gore. But scavengers hold both an ecological and capital value. In India, vultures are reported to be worth $10,000 apiece because they cut down disposal costs. In some mountainous regions of the world, particularly in the Himalayas, the rocky landscapes make for difficult grave digging, and the lack of trees means there's scarce fuel for cremation. A person's body is instead offered to the vultures and other scavengers in a "sky burial." To Tibetan Buddhist's, consciousness untethers from the body upon death. As dozens of vultures swarm the deceased like bees to a hive, a person's blood, flesh, and bone begin benefitting other living beings as a final act of compassion.

Scavengers help ranchers, too, by scrubbing pastures clean of dead livestock and afterbirth from calving. In mid-winter, a rancher near Stevensville, Montana, found his horse dead in an icy field. He figured it slipped and broke its neck. The frozen ground prevented him from digging a hole to bury it, so instead, he donated the carcass to Rob and Kate for their game camera project. He loaded the horse on a flatbed truck and drove it to a pasture across the road from Kate's house.

"The horse was huge," Kate said. "We had to have two cameras on it because there was no way to get the whole thing in a single camera frame." At the peak of scavenging, over thirty Bald Eagles were feeding on the horse. "I was so distracted," Kate said. "There would be eagles on the ground, on the horse, on the center pivot, on every post. They were like eagle lollipops. The horse was definitely more visible in the pasture from the main road than I had anticipated." She said that cars began pulling into her driveway to get a look at the eagles. Amish buggies even stopped.

Among the dozens of bald eagles that helped dispose of the horse, one was wearing a red leg band and a satellite transmitter. Kate learned that the Department of Defense had captured that eagle a decade ago near Flagstaff, Arizona. They never received additional details about the eagle's wanderings, but its mere presence 800 miles away from its initial capture location ten years later attests to the species' migratory nature and longevity.

In that first winter of the game camera project, over fifty golden and bald eagles wearing some form of auxiliary marker, from leg bands to wing tags, were recorded. That number doesn't reflect the dozens of unmarked eagles that scavenged. Kate and Rob were so bombarded with game camera photos, they were forced to recruit more volunteers. They ended up loading their images on Zooniverse.org, a community science platform that allows anyone with an internet connection to help researchers. In this case, thousands of armchair naturalists began classifying scavengers and looking for wing tags on eagles who were devouring roadkill (and a horse).

"One of the overarching themes of the eagle project was that people of all backgrounds, including ranchers, take interest and pleasure in the existence of eagles." Kate said. "Who would have thought that people with cows and sheep on their property would let us set dead deer there, knowing it would draw in potential predators? That's remarkable."

***

Scavengers haven't always been admired. In the late 1800s and early 1900s, ranchers and government-paid bounty hunters discovered that if they wanted to kill predators and scavengers, all they had to do was lace carcasses with poison. An eagle's versatility made the birds vulnerable. One minute, a golden eagle is bolting after a jackrabbit and a bald eagle is plucking a trout from a cold lake. If either bird sees a carcass the next minute, they might flip the switch to scavenger mode. That adaptability contrasts with vultures, who usually fuel themselves only with rotting animals. Yet, vultures offer a model for why eagles are easily poisoned by various substances, including lead.

Vultures find meat fast. In sub-Saharan Africa, their aerial circling casts an avian "Batman signal," alerting park rangers to a potential poaching event. Poachers would rather not get caught, so they pack carcasses with poison to rid the skies of vultures. Some African herders do the same, but they're battling lions and hyenas. Still, vultures die. All told, eight African vulture species have declined by an average of 62% over a thirty year span.

A worse die-off has played out across the Indian subcontinent. Sick and dying livestock were once administered a drug called diclofenac, a non-steroidal anti-inflammatory. The drug eased the discomfort of dying cows, but it spelled kidney failure for the vultures that later picked apart their bodies. The drug ushered in staggering declines of vultures, with some species losing more than 95% of their population. Without the birds polishing dead critters off alleyways and pastures, feral dogs filled their role. The dogs grew in numbers, started nipping more people, and eventually caused an increase in rabies deaths.

North America's vulture decimation story involves the California condor. During the Pleistocene Epoch that ended roughly 10,000 years ago, condors dined on the most enormous mammal carcasses ever to rot on North American soil, including mastodons, camels, and ground sloths. Condors ranged up and down the West Coast, across the Southern States, and along the Eastern Seaboard. After the Pleistocene's megafauna tromped their last steps, though, the condor's range shrunk to only the West Coast. In California, Spanish explorers witnessed condors tearing apart a beached whale. Others saw condors eating salmon in the Pacific Northwest and scavenging horses near Portland, Oregon. When the cattle industry expanded into California, condors feasted on dead beef.

Yet, as more humans settled within the condor's range, the birds were pushed toward extinction. In the early 1900s, the U.S. Government intensified its campaign against predators by hiring "hunters" to systematically poison wolves, coyotes, and other wildlife with fangs and claws. The hunters scattered cubes of fat across the landscape to habituate their future victims to a new food source. Afterward, they broadcast the fat cubes around bait laced with strychnine tablets. Sometimes they'd kill a horse on-site and pack it with the toxic tablets, ultimately causing scavengers to convulse until death.

Poisoning played a major role in the eradication of wolves from much of the West, while also slaughtering untold numbers of condors, eagles, and other opportunistic scavengers. By mid-century, the range of California condors had shrunk to only six counties outside of Los Angeles, California. The declines continued for decades, and by the 1980s, only twenty-two birds remained.

To save the species, biologists caught the survivors for captive breeding. The program was a success. In 1992, condors were released back into California and eventually into Arizona's Grand Canyon and Baja California, Mexico. Before telling the birds *adios*, the biologists established feeding stations that helped the condors transition into natural foragers. The stations also conditioned the birds to return to a capture location for wellness checks. That's how the biologists eventually learned that up to 95% of the condors had elevated blood lead levels, some condors requiring the same hospitalization and chelation therapy that Brooke Tanner administers at Wild Skies. The condors still faced other obstacles. They collided with utility lines, were electrocuted, and predators killed others. Chicks died when their parents fed them screws, bottle caps, and other micro trash mistaken as edibles. But lead poisoning posed the primary threat. The species would never sustain themselves in the wild if they required constant pit stops at the clinic.

Researchers were already pointing their fingers at spent ammunition. Nearly everyone knew that lead shotgun pellets poisoned birds, but the environmental aftermath of slinging lead bullets at animals remained unclear. A group of scientists from The Peregrine Fund teamed up with a veterinarian to give the topic a blast of evidence. The researchers collected thirty-eight deer that were shot by hunters in California and Wyoming, and then x-rayed the carcasses and gut piles. Bullet fragments lit up like constellations. Some of those particles were so small they could only be seen under magnification. In one deer, the researchers counted a startling 783 fragments.

Most lead bullets have a hard metal casing – usually copper or a copper alloy – that covers a soft lead core. They are designed to mushroom when they hit an animal. That expanded bullet tears apart tissue more violently than a bullet that keeps its shape. But when that soft metal flies out of a gun barrel at 3,000 feet per second and crashes into a warm body, lead particles splinter away and embed in tissues. Some fragments can end up more than a foot beyond the main wound channel. Hunters usually slice away bullet-torn tissues and toss them on the ground. The same goes for the internal organs. All of those tissues, plus the bullet fragments, stand good odds of sliding into a scavenger's belly.

Raptors can't fit much meat in their stomachs. Instead, they fill the crop in their throat, which acts like a grocery bag. In dangerous situations, such as when the golden eagle was whittling away the elk liver beside a skulking coyote, the bird was able to pack its crop quickly and depart unscathed. The eagle probably later digested its meal from the safety of a tree branch. Compared to the powerful gizzards of waterfowl and grouse that grind their food, raptors rely less on grinding and more on the acids that swish around their stomachs. It's generally true that animals who scavenge and encounter more food-borne pathogens have higher stomach acidity. A Shetland pony that eats grass will have a near-neutral stomach pH, like milk. If that pony dies, the scavengers that tear into its bloated body will probably have stomach pHs more acidic than lemon juice.

Stomach acidity has profound implications for lead absorption. Imagine lining up three empty shot glasses. Fill the first glass with tap water and plop in a hunk of lead. Lead is so insoluble in water, almost none of it will enter the solution. That's why when you cast a fishing line into a stream, your lead sinker doesn't dissolve. (Although lead sinkers and jigs do poison loons, swans, and other waterfowl when swallowed.) Now, fill the second glass with a strong acid and submerge another hunk of lead. The acid will start freeing lead into solution from the hunk of metal. Finally, fill the third glass with acid but before dropping in the lead, bust it up into fragments the size of beach sand and dust. By increasing the metal's surface area, more lead will mobilize into the solution. It's like how granulated sugar dissolves faster in tea than a sugar cube. This is what makes lead bullet fragments so dangerous to ingest.

With roughly 12 million hunters in the United States, there's no shortage of bullet fragments for scavengers to chomp between their beaks and teeth. Biologists have estimated that hunters leave 1.5 billion pounds of carrion in the field annually in the U.S. If all that flesh was ground into McDonald's quarter pound hamburgers, the patties would cover most of San Francisco. That remarkable pulse of grub largely coincides with the fall hunting season. It turns out that's when biologists saw lead concentrations spike in condor blood.

To really pin down if that lead originated from hunting bullets, biologists analyzed the isotopic fingerprint of the lead in condor blood. In 77% of the cases, it matched the isotopic ratio of the bullets sold locally. A follow-up study found evidence that condors probably snacked on lead-based paint from an old fire lookout, but lead ammunition was still the top source of exposure.

As condor biologists churned out evidence linking ammunition to lead exposure, interest ramped up to test other species, including humans. Researchers from the CDC tested the blood lead levels in residents from North Dakota who did or did not eat game meat. People who ate game meat *did* have higher concentrations of lead in their blood. Yet, the average blood lead level of game meat consumers was still lower than the national average. None of the study's participants had lead concentrations exceeding the CDC's level of concern. These results might put hunters at ease, but there's no known safe level of lead exposure, particularly for children. One study from Quebec, Canada, estimated that 7% of children who ate lead-harvested deer three times a week would be exposed to enough lead to decrease their IQ by one point.

University of Montana's master's student Thomas Rogers looked for lead in the large carnivores living in the Greater Yellowstone Ecosystem. Wolves and cougars usually came up clean. Meanwhile, black bears frequently had lead swimming through their blood, but at much lower concentrations than grizzlies. Yet, Roger's never found bullet fragments in grizzly scat, and bears sampled during the hunting season didn't test higher for lead. Rogers never pinpointed the source.

Particles of lead bullets didn't appear to be crippling the Yellowstone's iconic land carnivores, but they were weighing down the region's top aerial predators. Bryan Bedrosian, a co-author on Rogers' study and the biologist who suggested Rob test migrating golden eagles for lead, began sampling bald eagles near Jackson Hole, Wyoming. Each year, hunters killed about 3,000 big game animals in the National Elk Refuge, Grand Teton National Park, and surrounding areas. Out of the fifty-five bald eagles Bryan and his team caught during the hunting season, 98% of them had elevated blood lead levels. Like with the condor work, the spike in lead levels with eagles coincided with the hunting season, and none of the nestlings they caught in the spring and summer had elevated levels. (That uptick in the number of "leaded" eagles during and directly following the hunting season has been reported from the Pacific Northwest to Sweden.) As a final part of Bryan's study, his team equipped ten birds with satellite transmitters. Nine of them summered in Alberta, Saskatchewan, or Northwest Territories. When the birds migrated south, eight of them used the Jackson Hole area as a stopover site to feed on the glut of dead animals.

Similarly, about 90% of the eagles Raptor View had wrangled in the winter at MPG Ranch had elevated levels, compared to 58% in the fall on the Rocky Mountain Front. The eagle I held tested high, too. Still, I remained a steadfast lead slinger.

During my first semester of graduate school, classes were sapping my hopes of pursuing deer and elk in the backcountry. Instead, I set up a hunting blind at MPG Ranch near an alfalfa field. Within an hour, a whitetail doe meandered from a creekside cottonwood forest toward the crop, where I took my shot. I was thrilled to replenish my meat supply after consuming most of the previous season's elk. But guilt corroded the hunt. In six weeks,

Raptor View would be catching eagles 200 yards from where I carved the doe's organs from her steaming body. Where would those eagles take my bullet?

Many of the eagles in the Bitterroot Valley migrate. One January, Raptor View caught an adult bald eagle on private land where the property owners had participated in the game camera project. The biologists slipped a harness with a transmitter around the eagle's wings and cinched it at a leather breastplate. The setup resembled a small backpack. If it bothered the eagle, it could peck at the breastplate and the harness would fall off. They named the bird Relish, after a nearby landowner's locally famous zucchini relish.

Once out of the biologists' grip, Relish escaped to the Canadian Rockies and hung tight to the icy rivers that tumbled from the peaks southwest of Calgary. Within two weeks, Relish returned to western Montana and began bopping around the Flathead and Bitterroot Valleys. Come spring, she followed the spine of the Rockies and veered east over the Icefields Parkway between Banff and Jasper National Parks. She crossed into the Northwest Territories and settled along the Mackenzie River, 100 miles downstream from the outlet of the Great Slave Lake. She stayed put for six months, perhaps raising young where the boreal forest meets gray river water. Exactly one year after Raptor View fooled her with the net launcher, Relish returned to the capture site after completing a round trip distance of over 2,000 miles.

Golden eagles migrate farther. They cruise over the Rockies like Relish, but they stick to the mountain chain up to the Yukon Territories and Alaska. They nest on bluffs battling the cold winds that whip off glaciers. They fly to the Seward Peninsula, where the U.S. nearly kisses Russia and polar bears leave footprints on the coast. Come autumn, those eagles start spilling south from northern latitudes to their wintering grounds, where hunters like me are deciding what to leave in their paths.

6. THE COPPER OPTION

My breath rose like smoke under the glow of the headlamp, with my rifle slung on my shoulder. My friend, Tanner, dodged tree limbs as he followed behind. Dawn would soon light the first moments of the hunting season, but not until our legs burned from this absurd climb. We pushed between pines and spruce, navigated deadfall, and ascended to the elevation of subalpine fir. Tanner stood nearly a foot taller than me, with hip bones higher than my belly button, meaning he stepped over obstacles while I crawled. Branches caught on our backpacks and boot laces as we slithered through the brush. The brisk morning did little to cool our bodies, but the thought of bringing home an alpine whitetail motivated each step.

The trees thinned as we neared the basin, and we clicked off our headlamps. Deer scat lay on the dirt. I tapped the pellets with my toe. Dry and hard but still promising. I unslung my gun and cycled the action.

"I'm going to creep over and take a peek," I whispered. Tanner stayed back and watched me edge across the ridge. The sun hadn't crested the high peaks, but it offered enough light for the huckleberry bushes to radiate crimson. This mountain goat terrain seemed an odd habitat for whitetail deer, a species better known for hiding in woodlands and cornfields. Here, steep bedrock guarded the basin's highest ramparts. Below the cliff bands that framed its sides, avalanches had cleared out swaths of timber and left shrubby meadows. While exploring the area in July, I had spotted two nice bucks. But this morning as my binoculars brought the landscape's texture into view, those deer were either elsewhere or concealed. I opened the bolt and pushed the cartridge back into the magazine. Tanner caught up, and we continued farther over the ridge.

Back on the summer solstice, the tall stalks of beargrass here would have held bulb-like clusters of tiny white flowers. Now, thousands littered the ground like dry straws. Some balanced between clumps of grass. Some hid under shrubs. *Pop!* I stepped on a stalk. I turned to Tanner and cringed.

We tried stepping between grass stalks, but they were too dense and our feet too big. *Snap! Snap! Pop!* Below us, a group of deer startled from a clump of trees and began bounding away. Their white tails waved like metronomes through the basin's widest clearing. One was a buck, but even if he hit the brakes, he was too far for a shot. The deer veered downhill into thick timber. Gone.

"Well," I whispered. "That sucked."

"Sure did," Tanner replied.

We slipped off our packs and plopped onto the ground. I laid my rifle across my lap just in case another whitetail broke loose.

"Maybe they'll hole up in that forest," Tanner said, sliding a water bottle from his backpack.

I fished in my bag and unwrapped a protein bar. "One of us could drop down and loop through the trees," I said. "Whoever does it might get a shot." I gnawed the end off the bar and chewed. "If we don't get a shot there, a deer might spook into the clearing. One of us could hang here."

Tanner dug his heals into the dirt and leaned back. "I'll go down."

"You sure?" I asked. He was giving me the better deal.

"Yeah." He replied. "Where should we meet?"

"How about on the other side of those trees?" I said, pointing to a patch of timber across the basin. "I'll creep down the hill, too. That'll give me a closer shot if a deer pops out."

Tanner tucked away his water bottle and stood. He brushed dirt from his camo pants, shouldered his pack, and grabbed his gun. "See you in a bit," he said.

Tanner contoured down the open ridge. Beargrass stalks popped and snapped, and his pants swished against their flat leaves. *Damn beargrass.* He broke into the forest with branches brushing across his arms. His progress turned silent.

I descended, too, putting myself about 200 yards from where the deer had erupted into view. I nestled my rump on the soil, extended the rifle's bipod, and aimed at a branch downrange. My crosshairs seemed steady, so I chambered a round.

I had a renewed trust in my gun. After it failed to fire on my elk hunt, I brought it to a gunsmith. He nodded as I described the problem, then clutched the rifle with calloused hands and removed the bolt. Reaching under the counter, he retrieved a small tool and unscrewed the firing pin. The pin looked like a small chopstick. "You see this oil?" He asked, lifting it between us. "It's old." He barked. "When old oil gets cold, it turns to glue. Give me a minute."

He walked into a backroom with the firing pin and bolt. I heard him spray something from an aerosol can. Before I could take a lap around his tiny shop, he marched back to the counter. "I cleaned and oiled the pin." He slid the already assembled bolt into the gun, pointed the muzzle at a wall, and pulled the trigger. *Snap!* "Put the rifle in a chest freezer overnight. Shoot it in the morning. If it doesn't fire, bring it back. But I think it's fixed."

While waiting in the basin, the plan Tanner and I had hashed suddenly coalesced. A four-point buck jolted from the highest patch of timber and began hurdling off. I raised my rifle and bleated, making the low, guttural sound of a deer. Curiosity overwhelmed the buck and he stopped. The shot angle was too steep to use the bipod, so I lifted the gun and rested my forearm on my knee. Inhale. Exhale. Squeeze. *Crack!* The buck charged downhill and hooked around the timber.

"Damn," I muttered.

I ejected the round and loaded the next. A blaze orange vest flashed through the timber down to my right, and Tanner emerged. I grabbed my gear and started downhill. What was I thinking by taking such a long attempt off my knee? This wasn't a John Wayne western.

Tanner watched me descend the slope with frantic bounds. "What happened?" He asked.

"A buck jumped from the trees," I said, pointing uphill. "I didn't feel good about the shot."

"Where was he standing?"

"Farther down the clearing," I said. We hiked closer. I hunched over and shuffled through the area, looking for sign, scrutinizing each patch of earth and each blade of grass. "I don't see blood."

"If you're unsure about your shot," Tanner started, "we should probably give it a half-hour or so."

"Yeah," I nodded. "That's a good idea. Let's wait."

Tanner dropped his backpack and reached inside. He removed a gas canister and a small stove. "You want some coffee?" he asked, already sitting and stretching out his legs.

"No. Thanks," I replied. "I'm fine."

Tanner's dad had guided hunters in Colorado for a decade. I'm sure Tanner heard cautionary tales of hunters following poorly hit animals too soon, only to push their quarry past the blood trail and beyond recovery. My dad recounted similar stories. That advice had seemed obvious on car rides and around campfires, but a trigger pull lasts forever, and the desperation it can cause often clouds good judgment.

Tanner flicked his lighter and the stove hissed. I paced, rethinking my shot. And then there was my bullet – a copper bullet. I had driven two of them through a bull elk's lungs at MPG Ranch the year before during my final year of graduate school. The bull died where he clipped his last grass. But I hadn't shot a deer with one. Some hunters argued that they were lousy, especially for thin-skinned animals. Other hunters swore by them for all big game.

***

One of the best descriptions I've read of how copper bullets function came from Jack Carr's fictional thriller *The Terminal List*. The main character and former Navy Seal, James Reece, is betrayed while on duty overseas. The author, also a former Navy Seal, builds credibility by detailing the gear Reece uses in his path toward vengeance. In an opening scene, Reece climbs a hill outside Jackson, Wyoming, decked out in expensive Sitka camo to snipe one of his betrayers. Of all the bullets his character could have chambered, he chose the same bullet I fired at my "alpine" whitetail. Carr wrote, "The monolithic bullet was a Barnes Triple Shock, made from solid copper and scored inside the tiny hollow point to split into four petals upon impact like a deadly flower. It was engineered to penetrate deeply on big game animals and worked so well that special operations troops adopted it for use during the Global War on Terror."

Unlike lead bullets, most copper bullets rarely fragment once they hit hair. This means few if any pieces of metal end up on the menu for scavengers. There are exceptions. Cutting Edge Bullets manufactures a bullet coincidentally dubbed the "Raptor." The front of the projectile splinters on impact to throw six petals from the main wound channel to cause massive trauma. Yet, even if a scavenger, such as a raptor, swallows a copper fragment, that piece of metal will be far less toxic than lead. In one study, biologists fed captive American kestrels copper pellets nine times over thirty-eight days. The dosed birds didn't show signs of toxicity, and their blood copper concentrations weren't different from kestrels not administered the pellets. It's unclear if these results would transfer to other raptor species, but compared to the 1981 study where the bald eagles dosed with lead died or went blind and required euthanization, copper doesn't seem too bad.

Every hunter at MPG Ranch switched to copper. Hunting with lead and then watching Rob catch poisoned eagles didn't fit well with our organization's conservation goals. After shooting the whitetail doe with a lead bullet beside the alfalfa field, I wandered the aisles of a local sporting goods store, searching for copper bullets that matched my rifle's caliber. I located a small box of 50 Barnes TSX bullets—not ammunition—but bullets. That meant they had to be handloaded, which I wasn't equipped to do. I went home, searched online, and quickly bought two boxes of loaded ammunition.

Bullet experts trust copper, too. In Philip Massaro's book, *Understanding Ballistics*, he writes about the deadliest cartridges for squirrels up to elephants. When covering big-game cartridges, he often describes copper bullets as Carr did, but he adds that they're strong and retain their weight. High weight retention means maximum penetration. In an interview with RifleShooter magazine, the former CEO of Barnes Bullets, Randy Brookes, said, "If the bullet stops inside the body, it didn't cause maximum damage."

In 1979, Brookes decided to build a solid copper bullet after hunting brown bears in Alaska. Six years later, he killed a brown bear with one of those bullets. Brookes called the projectile the "X bullet," referring to the shape the front of the bullet makes upon upset. A year later, the product hit the retail market. Over the next thirty-five years, Barnes became a leading manufacturer of copper bullets, building a reputation for their bullets' superb precision and "whirling, destructive power." Barnes's focus was to build the best bullet; the fact that the projectiles didn't poison scavengers was a fortunate coincidence.

Lead poisoning in California condors brought on a more widespread use of copper bullets, commonly dubbed "non-lead." Incentive programs began for Arizona hunters to use copper. In two hunting areas within the California condor's range, the Arizona Game and Fish offered coupons for free non-lead ammunition to hunters who held big game tags. A follow-up survey showed a mostly positive perception of the copper option. For hunters who harvested an animal, 93% of them said copper performed as well or better than lead. Only 4% reported below average or poor accuracy and wouldn't recommend it to others. Although most hunters were satisfied with non-lead's performance, some had concerns over cost. Back then, a box of non-lead would have run about twice as much as inexpensive lead ammunition. (Today, Remington, Winchester, Nosler, Hornady, and other ammunition manufacturers construct and load copper bullets, with comparable prices to mid-level to high-end lead ammunition.)

In that 2006 survey, hunters voiced their opinions about the copper bullets. There were some gems:

"Accuracy was terrible," said one reviewer.

"The most accurate ammo I have ever used in my 30.06," said another.

"Penetration was not as good as lead."

"The bullet performed well, good penetration."

"Clean pass-through with no mushrooming."

"It made a bigger hole than lead and destroyed too much meat."

These handpicked contradictory statements reflect essential points. First, accuracy varies depending on the gun and the shooter. This holds true whether a hunter shoots lead or copper. If one box of ammunition doesn't shoot well, try another. Second, unless you have a background in crime forensics, it's tough to know exactly how your bullet behaved as it carved through the animal at supersonic speed. Take this statement from the 2006 survey: "Clean pass through with no mushrooming." This hunter probably reached their conclusion by studying the size of the exit hole. In reality, the only way to know whether a bullet mushroomed is to find the bullet. Expanded bullets for many calibers will only measure about as wide as a dime, regardless of whether it's lead or copper. And exit holes don't tell the whole story. Bullets crush, tear, and separate tissue in complex ways. Plus, shot distances vary, so does wind, altitude, point of impact, and the tissue the bullet encounters. A bullet that hits a shoulder blade can't be compared to a bullet that passes through soft tissue. With all these factors mingling in a microsecond, it's unfair to judge a bullet's performance based on a single kill. Even for a seasoned hunter who kills a couple of animals each year, coming to conclusions about individual bullets is tricky. Luckily, researchers have generated a healthy pile of studies showing copper penetrates, expands, kills, and flies as accurately as lead under their test conditions.

One study came out of Theodore Roosevelt National Park in North Dakota. Elk numbers had creeped beyond the threshold the landscape could sustain. When this happened in 1993 and 2000, the Park Service shipped off elk to Native tribes, government agencies, and other entities. But in the following years, fears of disease transmission derailed hopes of translocation, so the Park Service decided to cull a portion of the elk herd. Over four years, staff and volunteers shot 1,000 elk – all with copper bullets – and recovered 983 of them. The program donated roughly 150,000 pounds of "unleaded" venison to food banks, folks who helped, and Native American organizations.

A similar effort continues to play out at MPG Ranch. While living at the ranch, bands of elk trotted past my window on September nights, bugling, mewing, smashing fences, and raising hell like horny high schoolers. They raided our alfalfa fields, but that didn't bother us as much as their plucking thousands of seedlings from restoration areas. Then, once deep snow drove them out of the timbered high-country, the elk overgrazed the few native grass communities that had somehow survived decades of cattle stomping and grazing. The overabundant elk herd was jeopardizing its future habitat, so we began harvesting an average of forty-five annually, all with copper bullets. MPG's recovery rate of elk nearly matches the rate from Theodore Roosevelt National Park.

***

When it came to Tanner and me following that whitetail, my mind wasn't swimming with apprehensions about the copper bullet. My anxiety had everything to do with shot placement.

"It's been about a half an hour." I said once Tanner swigged his last coffee. "Should we see if I hit this deer?"

"Let's do it," he said, stuffing his cook set in his pack.

I stepped beyond where my pacing had flattened the grass. "You're not going to believe this, Tanner," I chuckled. "There's blood right here. Five yards from where we were sitting."

Blood coated grass blades and woody stems. It soaked bare earth. The blood trail widened for fifty yards until it met the collapsed buck. The copper bullet had entered through the buck's back half and exited his chest. A bad shot indeed, but a wound that probably ended his life within a minute.

With the warm smell of blood wafting into our sinuses, and grizzlies once again roaming my mind, we started processing the buck. The bullet had tore through his insides, so we didn't field dress him by removing his interior. Doing so might expose the meat to unwanted bacteria. Instead, we used the gutless method. I ran my knife down the length of the buck's back and peeled off the hide. A thick layer of waxy fat coated his hindquarters. The sub-alpine zone had offered him plenty of calories. We cut away the legs, neck, loin, and tenderloin before flipping him over.

As we stripped muscle from the other side, Tanner spotted the first scavenger. "Look uphill," he whispered. A long-tailed weasel was stalking in, hopping over branches and bending around brush. Its white coat arrived before the first snow. The tube-shaped carnivore stopped under a fir bough and perked up its pink nose. I tossed it a chunk of gristle. The weasel paid no attention. Its small black eyes focused on the larger gift we were about to leave.

7. SMALL GAME,
BIG EXPOSURE

"I'm looking for dead gophers," I said on the phone from my office.

"You're…uh," the rancher paused. "You're doing what?"

"It's an odd request, I know," I'd say, feigning a chuckle. "I'm running a scavenger study. We're trying to figure out what eats ground squirrels after they're shot. I'm not looking to shoot them. I'm just looking for ranchers who already shoot them. Then afterward, I'll set up game cameras to see which scavengers come in."

The length of the rancher's pause directly correlated with his degree of skepticism. "Who are you with again?" the man asked. Other ranchers shared his apprehension. After I was apparently too curious about one ranching operation, the landowner accused me of asking a lot of "pretty specific questions." A rancher in the Blackfoot Valley asked if I carried a sidearm. Grizzlies had been loitering around her home. Until state biologists relocated the bears, I'd need to pack heat for my little science project.

A few ranchers seemed genuinely curious about which scavengers inhabited their properties. But by the spring, when shooters began killing gophers, the ranchers' schedules didn't mesh with mine, or I didn't hear back. Maybe I did seem suspicious. After all, my job title on MPG Ranch's website was "Environmental Scientist." So here I was, some "environmentalist," calling from Montana's hippy refuge of Missoula, with a phone that had an area code from the ultra-liberal city of Seattle, asking to outfit their property with surveillance cameras.

Not surprisingly, my best connections came by word of mouth. After two months of networking, I had eight ranches lined up. My coworker, Lorinda, helped with one property near her hometown in White Sulphur Springs, about forty-five minutes east of Helena. The land belonged to her high school friend, Vance. He and his family ran cattle across several thousand acres and irrigated a couple of fields. Each spring, gophers turned portions of those pastures and croplands to Swiss cheese, so shooting them was business as usual.

Lorinda tagged along to help and see her family. We left Missoula on a May morning when the trees in town were dressed in vibrant green. As we neared White Sulphur Springs, signs of spring faded. Being 1,800 feet higher in elevation than Missoula, bands of snow still clung to the north aspects below ridgelines. The cottonwoods stood bare except for their newly budding leaves. The valley floor, however, was filled with lush grass and grazing cattle. Northern harriers glided over furrow lines, hoping to snatch a skittering rodent. A red-tailed hawk spiraled above.

Vance lived beside the Smith River. The stream's green waters sloshed against a rock outcropping near his house. Vance exited his garage when we rolled up. He carried an aluminum can of ammunition for his .22 long rifle.

Vance hopped in my truck, and he caught up with Lorinda on local gossip as we drove to the pastures. The truck jarred side to side as its tires fell in gopher holes and climbed over dirt mounds. Ranchers often despise ground squirrels for this reason: they transform pastures into terrain parks. They excavate holes wide enough for a rancher to twist an ankle. When a badger tunnels after the ground squirrels, the holes widen and the mounds grow even taller. The rough topography creates a nightmare for farmers trying to seed their fields.

Ground squirrels can chomp into a farmer's bottom line, too. I once visited a farm in Idaho where a fleet of gophers scurried from burrow to furrow, feasting on the seeds the farmer had just sown. Ravens formed gangs to slaughter the smallest squirrels. Bald eagles perched on the center pivot, waiting to make their play. But aerial predators couldn't stop the march of the squirrels, so the landowner rounded up shooters (and eventually poison) to help rescue his crop.

Vance wasn't battling such an epic swarm of gophers, but he still managed to thump a couple dozen before dinner time. Lorinda and I searched for the aftermath. At each carcass, I hammered a stake into the soil and strapped a game camera to it. Lorinda logged GPS coordinates and wrote notes. My truck rattled over freshly seeded fields toward a fence where Vance had dispatched two gophers. A pair of red-tailed hawks stood in front of the barbed wire.

"They're gonna steal the gophers before we set up the cameras," I said.

One hawk leapt into flight with empty talons. The second raptor lurched skyward but something was weighing it down. When the hawk finally lifted, we the saw the fat ground squirrel gripped in its foot. The bird pumped its wings with barely enough force to clear the fence.

"How do you account for that in your study, Mike?" Lorinda laughed.

As I watched the red-tailed hawk land 100 yards off with its lunch, I wondered if this raptor-rich environment would repeat the hawk show from a month earlier in Dillon, Montana. The most active game camera photographed at least five different hawks scavenging one gopher carcass, each pecking a few bites before flying off. Northern harriers arrived first, followed by two Swainson's hawks, which migrated all the way from Argentina to copulate on the carcass. (The photo later killed as a punchline in my presentations.) Finally, a red-tailed hawk took to the sky with the remains.

Vance, Lorinda, and I saddled into my rig and drove to a narrow valley. A ditch flowed through the middle, feeding willows and marshy soils. Terraces rose on both sides of the valley to benches above. Gopher mounds pockmarked the grassy incline. As Vance began his damage control, I noticed a golden eagle gliding above a rodent colony on the upper lip of the terrace, nearly brushing its wingtips across the grass. If that golden had a nestling nearby, I thought, the adult could be packing the nest with ground squirrels.

***

During the golden eagle breeding season in the western United States, rabbits and hares probably feed eagles more than any other prey. But in habitats where those prey are scarce, medium-sized rodents, such as ground squirrels, prairie dogs, and marmots, often comprise a major chunk of their diets. In David Ellis's book, *Enter the Realm of the Golden Eagle*, he includes a photo where adult golden eagles had surrounded their nestling with more than thirty ground squirrel carcasses, lining them up like sausages at a deli counter.

A wildlife photographer at MPG Ranch located a golden eagle nest nearly fifty feet up a conifer, with two eaglets standing on their stick house. A golden eagle alighted to the nest with a ground squirrel clutched in its talons. Instead of offering the nestlings bits of squirrel meat, one nestling clamped down with its beak, tilted back, and swallowed the entire rodent headfirst. The squirrel's tiny feet and furry tail slid out of view last, while the nestling's thin neck bulged like an overfed snake.

Biologists have estimated that shooters blast millions of ground squirrels, prairie dogs, marmots, coyotes, and other "varmints" each year.

Millions!

Plus, a shooter can kill dozens if not a hundred gophers and prairie dogs in a day, sometimes without a hunting license required. Unlike most big game animals that require the hunter to salvage its meat, varmints usually lay in the field, awaiting the tastebuds of scavengers. (Although, some online recipes claim prairie dogs make tasty pies and hot wings.)

In a 2006 study published in *The Journal of Wildlife Management*, researchers found that 93% of ground squirrels killed with lead .22 long rifle bullets retained fragments. About 20% of the carcasses were loaded with enough lead to potentially kill a raptor.

The .22 long rifle is the most popular cartridge in the world, firing a bullet about the size of a pencil eraser. It's as versatile to a small game hunter as a chef's knife is to a cook. Yet, the cartridge is slow, usually sending projectiles only a hair faster than the speed of sound, which clocks in at about 1,100 feet per second. Bullets fragment more violently the faster they fly, so the .22 long rifle's projectiles might actually hold together better than bullets shot from other cartridges, such as the lightning fast .223 Rem. The .223 pushes a slightly heavier bullet than the .22 long rifle, but at roughly three times the velocity. The bullet's extra velocity helps it fly with a flat trajectory and boosts the bullet's kinetic energy, vaporizing gophers into a pink mist upon impact. That explosive phenomenon perhaps explains why certain bullets are marketed with names like "Varmint Grenade." But speed comes at a cost. Biologists from the University of Wyoming x-rayed prairie dogs killed with the .223 and saw a spray of fragments. Some carcasses were littered with over a hundred tiny particles.

As I was finishing revisions on my master's thesis about the shooting range, MPG Ranch's general manager, Philip Ramsey, called me. His friend had shot fifteen gophers, and Philip asked a veterinarian to x-ray each one. Philip texted me a radiograph of a carcass speckled with fragments.

"We can find out which lead bullets are better than others," Philip said. "All the bullets won't fragment the same." The Wyoming study had already confirmed Philip's statement: expanding bullets left behind more lead than bullets that didn't mushroom. Yet, that study only included two types of ammunition—we could scale it up.

"I'd love to take the reins on a project like this," I said. "Maybe in the future we should toss in non-lead bullets. Things are going that way, and it might give the study extra appeal."

We ended up collecting over a hundred ground squirrels that were shot with one of eight types of ammunition. No matter what type of lead bullet a shooter fired, we spotted fragments behind the skin of nearly every gopher. And Philip was right, not all ammunition posed the same risk. Bullets with slower velocities tended to leave behind less shrapnel. Best of all, non-lead bullets dispatched gophers as swiftly as lead but without the toxic mess.

After the *Wildlife Society Bulletin* published our study, an editor from the organization contacted me for an interview. Lead poisoning was a pet issue for many biologists, so after the editor's article hit their website, emails poured in from federal employees, professors, and staff from non-profits and consulting agencies.

I used the moment to network, asking each biologist if they knew someone I might enjoy chatting with about lead poisoning. That strategy bounced my conversations from expert to expert across the nation. Everyone wanted to see non-lead outreach efforts ratchet up. A broad slice of the hunting community still didn't know their hunting bullets might sicken wildlife. A biologist at Bemidji State University in Minnesota and a future collaborator, Dr. Brian Hiller, kindled the biggest spark for a new study. "Maybe someone should watch dead ground squirrels with game cameras."

***

Lorinda and I returned for the cameras two days later as hawks whirled below storm clouds. I navigated my truck between cattle while Lorinda guided me to cameras with the GPS. All sixteen carcasses that had laid on the soil were gone, leaving only blood-soaked earth. The two carcasses that remained were the ones that had tumbled down burrows, hidden from birds.

I downloaded the photos on my laptop, expecting to see a magnificent raptor show, with eagles mopping up the remains, red-tailed hawks clutching gophers by the skull, and harriers excavating organs the way robins yank worms from lawns. What I saw stunned me. Ravens had pecked away most of the carcasses within hours of us leaving the cameras. I didn't recall seeing a single raven on the ranch.

Ravens are brilliant birds. With a brain the size of a walnut, they've performed as well as chimps on intelligence tests. They use sticks as tools. They trick other ravens by pretending to cache food in one spot but hiding it elsewhere. They are also the only animal experimentally shown to be attracted to gunshots. Some hunters claim that a gunshot rings the dinner bell for grizzly bears, but I'm unaware of a researcher reckless enough to test that theory.

The biologist who performed the gunshot study with ravens, Dr. Crow White (bonus points for the birdy name), visited a popular hunting area near Jackson Hole, Wyoming, and counted ravens before and after he shot his rifle. To make sure it wasn't just the gunshot or his presence that attracted ravens, he also blasted an air horn, blew a whistle, or did nothing at all. Only the gunshot lured ravens. Ravens are known to associate with predators to steal bites from fresh kills. Shooters are just another predator.

The cameras at Vance's ranch caught golden eagles on three occasions, but only after a raven or a magpie had landed. The corvids probably attracted the eagles. When one raven found a gopher carcass, sometimes a second raven landed. And then another. In between jabbing away flesh with their sharp beaks, they croaked. Then more ravens landed. Within ten minutes of a raven landing at one carcass, eleven ravens surrounded the gopher. Sixteen minutes after that, a golden eagle touched down. Ecologists call this relationship local enhancement. That's when the behavior of one individual draws the attention of other individuals. A golden eagle can't ignore a gang of squawking ravens.

To the ravens, eagles, and other scavengers, those dead ground squirrels create an ecological double-edged sword. Shooting the varmints creates a pulse of easy food that has the potential to enrich the nutrition of the scavengers and their offspring. In fact, a study from the Pacific Northwest showed that golden eagle nestlings gained more weight the closer they were situated to agricultural fields. Yet, the nestlings also had higher concentrations of lead in their blood.

***

In eastern Montana, locals shoot prairie dogs like gardeners pull dandelions. A friend in Miles City had connected me with two landowners in the area. My wife's family, who had deep roots near the North Dakota border, coordinated a third study site. A thunderstorm had whipped across the prairie the previous night, prying limbs from cottonwoods and flooding roads. I waited to meet the first shooter, Nate, until the sun dried our route. We met at the gas station in Terry, and by the early afternoon, we were driving beside railroad tracks and the Yellowstone River. The other direction, a narrow grassland slammed against splintered badlands streaked with oranges and browns.

We climbed a driveway and parked at a well-kept ranch house. Eager to head afield, Nate grabbed his shooting supplies, marched uphill to a dog town, and began his varmint control. He was firing a .17 Hornet, a cartridge that delivers a projectile the size of an infant's fingernail three times the speed of sound. After the Hornet had stung a couple of prairie dogs, the burrowers began hunkering down. But Nate knew their curiosity would eventually draw them into his crosshairs, so he paused his barrage. A prairie dog inched its head over the summit of its mound. *Crack!*

After twenty minutes of Nate shooting increasingly distant targets, the colony went on lockdown. Nate began exploring the outskirts of the dog town for any varmints that hadn't received the shelter-in-place signal. Meanwhile, I hiked out to the first colony with an armload of stakes, a sledgehammer, and game cameras strung across my chest like bandoliers. A graveyard of sun-bleached prairie dog skulls and leg bones rested among the freshly deceased. The remains attested to the volume of shooting the landowners allowed. I deployed my cameras, returned to my truck for another load, and found Nate at the far edge of the dog town.

"What do you think the cameras will catch?" Nate asked.

"I don't know," I said. "So far this year, we've seen badgers, coyotes, lots of raptors. Who knows, maybe an eagle."

"We saw a tagged eagle here a couple years ago," Nate said. He removed his phone from his pocket and slid through photos, showing me a golden eagle he had watched feed on a dead cow beside the Yellowstone River. The bird wore blue tags with white numbers.

"I know who tagged it!" I said.

Nate shuffled through more photos and held out his phone. "Was it this guy?" The man holding the golden eagle was Heiko Langner, a former geoscientist at the University of Montana. He worked in my department while I was in graduate school, and he co-authored Rob's study about lead exposure in migratory golden eagles. Raptor View had caught this particular eagle as a first-year bird on the Rocky Mountain Front, ten years before Nate took that photo.

My game cameras didn't capture that eagle or any eagle in Terry. Instead, they revealed a first: burrowing owls. These raptors look like feathery softballs standing on stilts. They typically eat small critters, such as bugs and lizards, but these owls were scavenging prairie dogs pre-seasoned with lead. It seemed every opportunistic meat-eater relished the opportunity to shred a dead varmint into bite-sized pieces.

Condors and eagles draw the most attention as being victims of lead poisonings, but at the eight ranches we sampled, more than a dozen other species scavenged, like burrowing owls. How many of these opportunistic scavengers die or become debilitated by lead but their weakness goes unnoticed?

Before driving back to Missoula, I cruised the aisles of a sporting goods store in Miles City. The shelves were stacked deep with lead ammunition. Their non-lead offerings were scant and limited to big-game cartridges. Even though Philip and I had published a study in a wildlife journal touting the effectiveness of non-lead varmint rounds, their availability was poor everywhere besides at online retailers. Many small-game hunters probably didn't realize they had an alternative. We needed to ratchet up outreach.

8. LEGISLATE OR EDUCATE?

The gulley that divided the grassland blocked the breeze, but the April air was still chilly. The low, white clouds threatened snow. I hoped they'd wait. A journalist had just run a story in the local newspaper about this shooting demonstration at MPG Ranch, where we were promoting copper bullets instead of lead. I wanted a good turnout. About twenty-five people had already arrived. They were milling around the gravelly ravine, some talking, others kicking their toes in the rocks, all waiting for it to start. I saw a latecomer walking up the gully from the parking area. He had a blonde mustache and was wearing a brown Carhartt jacket. I held off introducing our guests so I could greet him.

"Thanks for coming," I said. "How did you hear about the demo?"

"The paper," he replied, raising a folded copy of the Ravalli Republic. His chin began quivering. "Are you for this or against this?"

"Uh–," I tried to respond.

"People are using copper bullets, and animals are running off," he said with a pointed finger. "The bullets don't work."

"Well ..."

"I've been using lead all my life," he continued, becoming more aggressive. He pulled a plastic bag from the inside pocket of his jacket and held it at eye level. Three muzzleloader bullets hung heavy at the bottom. "Are you going to tell me *these* fragmented on elk?"

I felt pinned against a wall. I had only seen people react this viciously to non-lead ammunition in online forums, but never in person. This man was passionate about the hunk of metal that left his gun barrel. He needed to vent, and he seemed to be gaining momentum. Instead of pushing back, I walked toward the presenters. "The demo is going to start," I said over my shoulder. "Let's see what these guys have to say."

The presenters, Chris Parish from The Peregrine Fund and Leland Brown from the Oregon Zoo, were both hunters and biologists who thrived on heated conversations. Chris had helped run many of the early studies that highlighted lead poisoning in raptors, particularly in California condors. He was an ex-football player who filled out his brown overalls and wore a trucker hat on his bald head. Leland was tall, slender, and had a trimmed beard. The Oregon Zoo hired him to run a non-lead outreach program. Before that, he killed feral pigs and other invasive species in Hawaii and California for work, all with non-lead bullets. He knew their ballistics as well as anyone.

Chris began the demonstration by describing how he and Leland were helping promote ecosystem health by using copper bullets that rarely fragment. When Chris paused for a breath, the mustached man hollered out the claims from earlier. He showed everyone the muzzleloader bullets, and then he pulled a magazine article from a jacket pocket. "Birds are getting hit by wind turbines too!"

"You just posed a lot of questions," Chris said, "and I want to answer all of them. You, Sir, are why I am here. I'm Chris Parish and pleased to meet you. Your name is?" Chris's remarks caught the man off-guard. Then, Chris acknowledged that wind turbines were a threat to raptors, but so was lead ammunition. He justified how the demonstration would help us explore that idea.

"I think I need to apologize," the man said. "I might have come on a little strong—especially to this guy over there." He nodded at me. The crowd was baffled. Chris had disarmed him like a trained mediator.

Chris continued his introduction until he was ready to shoot. Everyone stuffed in their disposable earplugs and stood back. Chris set his .308 Windham assault rifle onto a gun rest. His target was a ballistics gel sitting fifty yards away in a rain barrel. Ballistics gels are clear, rectangular solids about the length of two shoe boxes. They feel like firm Jell-O, and the FBI uses them in forensic research because the gel mimics flesh. Leland watched the gel through a spotting scope as Chris's rifle sent a lead bullet downrange. The gun's report echoed off the gully's slopes. The gel heaved upward, flexing like an overfilled water balloon in midflight. It landed with a hard smack.

After firing a copper bullet at a second gel, Chris opened the rifle's action, stepped away from the shooting bench, and announced the range clear and cold. Everyone removed their earplugs and walked out to the gels. Chris pulled them from the barrels and set them on an adjacent table. The copper bullet had sliced a clean path through the gel, whereas the lead bullet had splintered and left behind tiny flecks of metal and dust. People leaned in to look from the side, studying the poisonous leftovers. They stood on their toes to see it from above. "Yuck," said one man. The man with the mustache didn't speak.

***

The shooting demonstration happened on a whim. I'd phoned Chris for an interview when he mentioned that he and Leland were about to be in Missoula, running a booth about non-lead bullets at the Backcountry Hunters & Anglers Rendezvous. When he offered to give our staff a shooting demonstration sometime, I asked: "How about next week?"

I was talking to Chris for an article I hoped to publish in a hunting magazine. The evidence linking big game hunting to lead exposure in wildlife was overwhelming. Yet, hunting and shooting organizations rarely talked about it. Meanwhile, websites tried to discredit the science or spread misinformation. I wanted to clear up the confusion. One pesky website called huntfortruth.org ignited my motivation more than any. They posted paragraphs like this:

The crux of anti-hunting activists' argument against traditional ammunition rests on the misplaced assertion that the use of lead ammunition for hunting leads to elevated lead exposure and poisoning in scavenging animals, such as the California condor, that allegedly ingest fragments of spent ammunition in gut-piles or carcasses left in the field by hunters. The scientific studies relied on by the anti-lead proponents are in fact not scientifically sound. In other words, the proponents use "faulty science" to support their anti-lead ammunition agenda. Huntfortruth.org has procured and analyzed over one hundred thousand documents from governmental agencies, universities and researchers and have found systemic flaws, which include faulty methodology and sampling protocols and the selective use of data (i.e. "cherry picking" data for publication).

Reading that paragraph set my mind whirling with counter-arguments, but after being steamrolled by the mustached man, I was unsure how to deliver those talking points in person. That's why watching Chris and Leland work the crowd at the shooting demonstration was the day's highlight. After Chris explained the ecological implications of hunting with lead bullets, Leland spoke about the ballistics of copper and how their qualities make them a premium bullet. Not only could hunters improve the health of scavengers, but they might also prefer the new bullet over lead. Chris and Leland were biologists with people skills. When a spectator threw a verbal punch, instead of punching back, they responded with the equivalent of: "Hey, nice punch. Let's modify your stance so you can hit even harder." A marriage counselor would have scribbled notes.

Those techniques were forged and honed from years of combative arguments. Websites and online forums had perpetuated the idea that promoting non-lead ammunition was spearheaded by anti-hunters. That hostility spawned from organizations petitioning against the use of lead ammunition. Then, legislators joined in without consulting the hunting community. The resulting division probably explains why I later encountered so much difficulty publishing my article in a hunting magazine.

By the early years of this century, studies had piled up indicating lead poisoning threatened the recovery of California condors. During the 2007-2008 California legislative session, a bill was introduced to enact the Ridley-Tree Condor Preservation Act, requiring big game and coyote hunters to shoot non-lead ammunition in the home ranges of California condors. At the time, Leland worked for the Institute of Wildlife Studies in California, controlling invasive species on the Channel Islands. In 2010, he switched jobs, becoming IWS's non-lead outreach coordinator. Leland said the previous guy in that position received death threats. For Leland's safety, the IWS kept his physical location a secret like he was enrolled in the Witness Protection Program.

At outreach events, hunters yelled and screamed at Leland, saying a ban on lead bullets was a step toward banning hunting. Leland reminded them he wasn't the one who passed the legislation. He was just helping hunters understand their bullet options. Still, some hunters tested Leland's composure by calling him anti-hunter, ignoring that Leland is not only a hunter, but he has promoted hunting his entire professional career.

"Basically, hunters went to bed one night being able to use lead bullets. They woke up the next morning not being able to use lead bullets. There wasn't a lot of outreach," said Russell Kuhlman, who took over soon after Leland left the position. Russell had previously taught active military personnel and their families how to shoot trap and skeet at Fort Riley, Kansas. While continuing Leland's past outreach work, Russell helped operate the huntingwithnonlead.org website. The site gives hunters tips on switching to non-lead ammunition and holds a repository of scientific literature. Russell also tested ballistics and reloaded ammunition, adding those findings to the website and building his repertoire of talking points.

In 2013, California edged closer to a statewide phase-out of lead ammunition by passing Assembly Bill 711. On July 1, 2015, non-lead would be required when hunting all California Department of Fish and Wildlife lands and where hunters pursued desert bighorn sheep. A year later, non-lead ammunition would be required when hunting with shotguns for small game and most upland birds. By July 1, 2019, hunters were required to use non-lead ammunition statewide for any purpose other than target shooting.

Among the many events Russell attended to run a non-lead booth, the biggest was the International Sportsman Expo in Sacramento. Between 30,000-50,000 people walked the aisles, book-ing guided hunts to Africa, hearing about the latest optics, and running their palms along the gunwales of jet boats. Each year, Russell and his colleagues tallied about 2,000 conversations over the four-day event. They considered it a litmus test on where the hunting community stood with non-lead ammunition. In 2014, few hunters entertained the idea that a soft lead bullet traveling at 3,000 feet per second could fragment and poison scavengers, so Russell heard fanciful conspiracy theories for how raptors were exposed to lead. Someone told him birds eat the lead that balances car wheels. Or, cars that hit deer dripped antifreeze on the carcass. Because antifreeze contains lead (it doesn't), condors ate lead by scavenging roadkill. Another person said airplane fuel was the culprit. Birds fly through the exhaust and breath the fumes. As ridiculous as this sounds, some smaller aircraft do still burn leaded fuel called "avgas." Yet, I've never met a pilot who watched an eagle follow their plane just to swallow miles of exhaust.

The year before California's ban took effect, Russell heard fewer conspiracy theories, and the fact that scavengers ate bits of lead bullets was old news. Hunters were more curious about reloading information. They wanted to hear about the new copper bullet advertised to expand at long-range. One guy even apologized to Russell, saying he had acted like a jerk the previous year. The guy had watched his buddy kill an elk at 300 yards with a copper bullet. He said it expanded and passed through the entire animal. Russell got the impression that hunters who switched to copper stuck with it, even when they didn't have to. They used it for Colorado elk, Wyoming mule deer, or whichever state and species they hunted next.

While California was phasing out lead for hunting, petitions were being drafted to ban lead ammunition at the national level. In 2010, the Center for Biological Diversity, the American Bird Conservancy, and several other groups petitioned the EPA "to ban lead shot, bullets, and fishing sinkers under the Toxic Substances

Control Act." The TSCA was passed in 1976 to give the EPA the authority to regulate chemical substances like lead. But the petition cast a wide net—target shooters wouldn't even be able to buy lead ammunition.

Critics often accuse the U.S. Government as being slow to act. Not here. The EPA denied the petition by the end of the month. Their letter included only four sentences, basically saying the TSCA doesn't authorize the EPA to address lead shot and bullets as requested. That setback didn't curtail the petitioners' fervor. They recruited more supporters, totaling 101 organizations, and tried again two years later. They were denied.

And then the Humane Society of the United States gave it a go. They gathered supporters and wrote a petition to the Department of the Interior in 2014, asking for nontoxic ammunition to be required on lands operated by the National Park and the U.S. Fish and Wildlife Services. That letter didn't result in immediate action limiting lead, but in 2016, during the last days of President Obama's second term, his administration ordered a phase-out of lead ammunition on national wildlife refuges. The move required the use of nontoxic ammunition and fishing tackle by January 2022. The order also directed the agency to work with federal, state, and tribal agencies to achieve that goal. About a month and a half later, President Trump's freshly sworn-in Secretary of the Interior, Ryan Zinke, reversed that order.

Any imposed restrictions on ammunition will probably be greeted with hostility, and outright bans may have pitfalls. Dr. Clinton Epps, a biologist at Oregon State University, published an article in a 2014 issue of the scientific journal *The Condor* that highlighted the practicalities of switching to non-lead ammunition and the potential challenges of a ban. It's one of my all-time favorite scientific articles, so I reached out to him like a groupie.

"I was initially sort of reluctant to get involved with this topic,"

Dr. Epps said on the phone. "But because I consider myself a conservation biologist and a wildlife biologist, as well as a hunter and a shooter, I felt that I had perspectives on both sides of the issue and that I might have a bit of a role to play."

In his article, Dr. Epps suggests that a ban on lead ammunition for hunting might not unfold like the lead ban for waterfowl hunting. For one, the logistics of enforcement may be trickier. Compared to waterfowl hunters that often concentrate on or near public waterways and wildlife refuges, big game hunters hunt a wide variety of public and private lands, frequently in lower densities, thus creating difficulties to wardens checking for compliance. Moreover, many lead bullets have a copper jacket and a non-metallic ballistic tip, making them indistinguishable from non-lead in a visual inspection.

A bigger challenge would be supplying hunters with ammunition. Big-game hunters shoot dozens if not hundreds of different rifle cartridges. Ten or more lead-free options may be available for common cartridges, but maybe not that old 6mm Remington your grandfather gifted you after graduation. Even if a hunter finds lead-free rounds for an uncommon rifle cartridge, its accuracy is not guaranteed. In comparison, waterfowlers use shotguns that come in only six gauges, with the 12 and 20 gauge being the most popular.

Dr. Epps argues that the best path forward requires "buy-in" from hunters. Otherwise, imposed restrictions on hunting ammunition could be perceived as unfair or an attack on second amendment rights. "My feeling is we would do a lot better to try to engage people with this on a voluntary basis," said Dr. Epps. "I'm afraid that if it's done with a legislative ban, there will be a pretty high number of people that don't comply, and then the net impact on the landscape may not change that much. I don't have evidence of that. That's my opinion, basically."

Despite California phasing out lead, a 2011 study showed that blood lead levels from a subset of condors had not decreased. This is possibly due to the birds becoming more independent to free forage and not relying on food supplements, or hunters might not have complied with the new regulations. Although repeating the study today might offer different results, the finding draws into question the effectiveness of a ban. The ban did, however, decrease the levels of lead in the blood of turkey vultures and golden eagles in the study area.

***

Meanwhile, Arizona Fish and Game avoided legislation and instead focused on education and outreach. Chris Parish and his colleagues at The Peregrine Fund had brought data about lead poisoning to their state game and fish partners. The Peregrine Fund showed the x-rays of deer shot with lead bullets. They shared location data from GPS-wearing condors and how the birds frequented the Kaibab Plateau during fall hunts—a behavior that coincided with spikes in their blood lead levels. Chris said the Arizona Game and Fish didn't shy away from the information. But before implementing any form of management, the state surveyed landowners, locals, and hunters to see what it would take to acquire their "buy in." In short, everyone wanted to see the data.

The agency produced a mailer with the condor and bullet fragmentation data, while generating an opportunity for hunters on the Kaibab Plateau to receive two free boxes of non-lead ammunition. Hunters who still shot lead but packed out their gut piles were entered into a raffle to win gift cards, fishing trips, and even an elk hunt on the Navajo Nation. About 50% of hunters participated. A good start, but not enough.

"The problem wasn't the information." Chris Parish said, who worked closely with Arizona Game and Fish to implement the program. "It was a three-page document with probably 8-point font. Some hunters weren't reading it. Who reads mailers?"

Arizona Game and Fish overhauled their approach. They produced a short film about lead poisoning in condors that was narrated by hall of fame pitcher Nolan Ryan. They then began slipping mini-DVDs of the film into the envelopes containing a hunter's tag. Participation shot up to 80% and has averaged an 87% annual participation in the dozen years since.

"Hunters have a long history and rich tradition of conservation," Chris said. "When you make an appeal to them and say, 'Here are the data, and here is what we're asking your help with,' hunters have always said, 'You bet, we can help.'"

***

Hunting organizations have not been fond of proposed bans, either. In response to the Humane Society's 2014 petition, thirty-three national sportsman's conservation organizations, including the Boone and Crockett Club, Rocky Mountain Elk Foundation, and Ducks Unlimited wrote a response to Sally Jewell, the Secretary of the Interior. They began by noting the Humane Society "is an avowed anti-hunting organization" and "one of its primary goals is to shut down all hunting." Later, they wrote "that if such a ban was instituted, additional petitions for the Bureau of Land Management (BLM) lands and other federal holdings would likely follow." They go on to say approximately 70% of ammunition purchases go toward non-hunting purposes, with "traditional ammunition" comprising 95% of all ammunition sales.

With copper ammunition being more expensive and unavailable for many cartridges at the time, if fewer people bought ammunition, the requested ban could decrease conservation dollars generated through the Pittman-Robertson program.

The letter raised one point that wildlife managers continue to grapple with:

The issue at hand is not whether lead, when ingested in a sufficient quantity and metabolized by an animal can be injurious to that par-

ticular animal, but, instead, what is the population-based impact to a species at a local or regional level.

Eagle populations in the U.S. aren't nosediving toward extinction. Bald eagle numbers are booming. They seem to burst from every riverside forest I explore.

Golden eagle populations are more complicated. The Western U.S. holds 30,000 individuals, migrants from Alaska add approximately 12,000 more each winter, and the Eastern U.S. supports about 5,000. Overall, their numbers are stable, according to a 2016 report by the U.S. Fish and Wildlife Service. Yet, zooming to specific regions can offer different perspectives. Rob and his trained crew of young eyes continue to count fewer migrants on the Rocky Mountain Front compared to when he started. Farther north in Alberta, biologists have recorded a similar downward trend, although those observations could indicate changes in migratory behavior rather than a shrinking population. As you descend in latitude toward the southern reaches of the western United States, golden eagle numbers appear to be creeping downward. Long term studies from California to Idaho have recorded 30-44% declines in nesting populations.

Chris Parish thinks a population-level decline is not needed before hunters should act. "Can we as hunters move into the future of hunting and defend our rights to do so if we are knowingly putting animals in harm's way because of our tradition of using lead-based ammunition? The problem isn't that hunters don't care, it's that they don't know all of the facts. That's where we can help by navigating the science."

9. AMERICAN EAGLE

The fly line unfurled off my rod tip, thrusting a streamer to the opposite bank. I yanked the line and my olive sculpin swam from the shallows to heavy current. A brown trout shimmered its buttery side and gulped, splashing its tail against the foamy water, fighting the straining rod.

In my hand, I let the trout wriggle free before fishing downstream to the tailout. My sandals gripped cobble as I waded across Rock Creek, just over the Sapphire Mountains from the Bitterroot Valley. At the other side, I grabbed a fistful of grass and pulled myself onto the floodplain. I stepped off the fisherman's trail and pushed through a thicket, knowing a shortcut to deep water under a cliff. My path opened to a patch of conifers, where a dead raptor lay beneath the pines.

The bird's body had settled flat as if its flesh had been vacuumed away, leaving only a bed of dark feathers and eight eagle-sized talons. I couldn't identify the bird as a golden or a juvenile bald—I was in college and hadn't yet undergone my birdy transformation. Today, I still contemplate what killed that raptor.

Eagles face a diversity of threats, both indirect and direct. Golden eagles are often squeezed out of quiet nesting territory by suburban sprawl and energy development. Wildfires have scoured shrubs from grasslands, eliminating habitat for one of their main prey species: jackrabbits. Goldens lucky enough to claim prime terrain might find trouble reaching old age. The odds a golden will survive its first year is 70%, mostly the result of starvation and disease. Golden eagles three years and older stand better odds, with an 87% chance of making it to the next year. These older eagles are seasoned hunters and foragers who have learned to fill their bellies year-round. They rarely die from natural causes unless it's from an injury, such as while fighting to protect a nesting territory. Their primary killer? Humans.

Given the huge distances some eagles travel, they constantly maneuver their lives through a deadly obstacle course. It's difficult to estimate how many eagles perish from each hazard. A bald eagle that collides with a semi-truck will more likely end up a statistic than an eagle poisoned in a quiet forest. Still, eagle conservation requires mitigating the major risks as best as possible and understanding how those risks interact.

Trapping

On July 6, 2015, near Nome, Alaska, Dr. Travis Booms, the biologist who spotted one of Rob's tagged eagles in Denali National Park, scrambled up a rocky slope to a golden eagle nest overlooking rolling tundra and a willow-lined creek. The white puff of a future predator watched Booms with more curiosity than panic. From the male nestling's rock ledge on the Seward Peninsula, the eaglet had likely spotted caribou and grizzly bears, but never a human. Booms carefully slid a metallic band etched with the number 0709-02908 around the eagle's ankle and tightened the loop with banding pliers. He didn't test the nestling's blood for lead, but the other nestlings he had tested nearby all had low concentrations, if any lead could be detected.

Nearly six months later in December, Raptor View caught a golden eagle on their bait at MPG Ranch, over 2,000 miles from the Seward Peninsula. The biologists felt a rush of excitement once they noticed the eagle was wearing a silver leg band. With such a rare catch, they wanted to equip it with a GPS transmitter, but they needed to ask permission from the original bander; otherwise, they might interfere with ongoing research. After a sequence of frantic phone calls, Rob Domenech had Booms on the phone, who was thrilled to learn about the recapture. He gave Rob the go-ahead to affix a transmitter to the eagle.

The biologists sampled the eagle's blood for lead. It was nearly four times the level considered elevated, but still below a debilitating concentration. Following the release, the biologists nicknamed the eagle Yuri, after Yuri Gagarin, the Russian cosmonaut famed for being the first human in space.

For the next month and a half, Yuri hunted and scavenged among snow-laden trees as he flew over northwest Montana and into Idaho. He began hop-scotching over the Canadian border for the rest of winter, just south of Kootenay Lake near Creston, British Columbia, where Kokanee beer is brewed. In April, Yuri flew northeast toward Banff National Park before tightening his movements northbound. Flying at a rate of about 130 miles per day, Yuri arrived in Alaska's interior within two weeks, passing for-ty miles from Booms' office in Fairbanks.

Yuri explored the Last Frontier all summer. He nearly reached the Seward Peninsula where Booms banded him, but instead of hunting where he'd spent his first months of life, he banked northeast to the Gates of the Arctic National Park and Preserve. Yuri toured the southern ramparts of the Brooks Range, past wind-rippled lakes and over the Trans-Alaska Pipeline System. He spent three weeks hunting the lush tundra near the remote community of Arctic Village, nestled on the edge of the highly politicized, oil-rich, and caribou dense, Arctic National Wildlife Refuge. Yuri then lifted skyward and traversed the Brooks Range, cruising within twenty miles of the Arctic Ocean and less than seventy-five miles from Barrow, the northernmost town in the United States. Yuri scouted the North Slope for three months, undoubtedly plunging his talons at willow ptarmigans and arctic ground squirrels. In September, he cut a path south again through the Yukon Territories toward his winter home near Creston, Brit-ish Columbia.

Over the coming years, Yuri migrated to Alaska each spring but focused his travels on the Seward Peninsula, where he had hatched. During his fourth autumn, Yuri followed his now regu-lar route south between the Coast Mountains of British Colum-bia and the Rockies. In early December, near Fort Saint James in central British Columbia, Yuri died in an island of timber among clear-cut forest after being caught in a snare trap set for wolves.

Remarkably, the trapper who killed Yuri accidentally killed anoth-er golden eagle that Travis Booms had banded. Booms had caught the second eagle and outfitted it with a GPS transmitter during spring migration on Gunsight Mountain, northeast of Anchorage, Alaska. Only a small fraction of the world's golden eagles fly with transmitters, so the odds of the wolf trapper catching two of them, both handled by Booms, were extraordinary. To the trapper's credit, he reported both mortalities. When they spoke on the phone, Booms recalled the trapper seeming genuinely upset about killing the birds. Still, wherever there's a hunk of meat, an eagle might find it, and if there's a trap nearby, an eagle might find that, too.

Yuri's demise in a trap was unfortunately routine for eagles. Biologists all too often cringe while spotting eagles cruising overhead with broken snares around their necks. Game cameras have photographed eagles with leg hold traps clinging to a foot. Raptor View regularly catches eagles with wounded feet. Rob Domenech and his team send the worst-off eagles to Brooke Tanner at Wild Skies, who treats those birds for infection and trauma.

Catching an eagle is probably one of the last things a trapper wants (unless they're illegally selling eagle parts). If they find an uninjured eagle tangled in their set, they must release it carefully so a talon doesn't skewer their hand. If the eagle is in bad shape or dead, the trapper is usually required to bring a wildlife official to the site, depending on the local regulations. Either way, it's stressful if not deadly to the eagle and a hassle for the trapper.

State and provincial trapping regulations vary, but they typically have guidelines to help prevent incidental catch. In Montana, for example, it's illegal to set a trap within thirty feet of a carcass or bait, unless the bait is concealed from above. Most raptors rely on their keen eyesight to find prey and carrion, so if they can't see the bait, they might not investigate (at least that's the theory.)

In reality, raptors can locate food by cuing into the behavior of other species. If an eagle sees a Steller's jay extracting nuggets of meat from under fir boughs, the eagle might touch down for a look. And placing the trap away from the bait won't necessarily shield eagles from accidents. Eagles often land and hop toward a carcass, so a trap could easily intercept them mid-bounce. If they jump atop the bait unscathed, a competing eagle or coyote might invite a skirmish, stirring up frantic footwork and wing flaps, boosting everyone's odds of stepping on a leg hold trap or getting strangled by a snare like Yuri.

Collision

The oldest recorded bald eagle was at least thirty-eight years old when a car smacked it in New York. Vehicles kill heaps of eagles. When snow piles up in the high county, deer, elk, and other species often migrate to valley bottoms, sometimes near highways. When an animal collides with a car hood, spilling blood across the pavement, the carcass becomes a hazardous attraction to scavengers. In a single winter on the highways near Rock Springs, Wyoming, cars collided with nearly 100 golden eagles.

And then there's wind farms. At their best, they generate renewable energy that help curb greenhouse gas emissions. At their worst, their spinning blades form a field of aerial meat grinders. Although wind operators are rapidly developing techniques to prevent wildlife fatalities, wind farms still kill somewhere between a hundred thousand to a half-million birds of various species annually in the United States. Golden eagles are particularly at risk because they seek out open expanses that offer updrafts. These areas also make ideal wind farm habitat. Eagles don't cluelessly glide toward a turbine like a fly to a bug zapper. Instead, they're busy hurtling from the sky at 200 miles per hour chasing prey and jousting with competitors mid-air. These aerial undertakings could easily distract them from a blur of turbine blades slicing upon their heads at over 100 miles per hour.

Electrocution and Shooting

Raptor View's 4Runner's sped by me as I walked the gravel road at MPG Ranch. The vehicle curved up to a house. I followed, assuming they had an eagle on board.

At the house, one of Rob's biologists, Shane, held an adult bald eagle by the legs. The bird's chest muscles had atrophied and were sunken toward its ribs. Lice crawled between its feathers and began marching up Shane's sleeves like minuscule army ants. The tip of the bird's beak was overgrown and askew as if you'd moved your jaw a half inch to the side. The feathers on the upper ends of its legs were singed, probably from being electrocuted. To round things out, the eagle had elevated lead levels.

Rob wanted a second opinion, so he called Kate Davis, a raptor expert and falconer who lived down the road. Meanwhile, Raptor View took their normal measurements and slipped a metallic band around its leg.

After Kate pulled up in her Subaru, she inspected its deformed bill like a dentist looking for cavities. She pulled toenail clippers and a fingernail file from her faded Carhartt jacket. With a few snaps of the clippers and sweeps of the file, the bird's beak regained a more normal shape.

Rob slid the falconry hood off the eagle's head, and the bird's wings flared and its beak gaped. The eagle still had fight. Kate dusted the bird with delousing powder, and afterward, Shane fed the eagle nearly a pound of venison. When Rob tossed the eagle toward the horizon, everyone wondered whether the bird would survive the winter.

Nearly two years later, the eagle turned up near Creston, British Columbia (the same Creston where Yuri spent his winters). Supposedly, the eagle attacked a lamb, so the livestock owner shot the bird dead. A local biologist recovered the eagle and relayed the band number to the Bird Banding Laboratory before contacting Rob, who never learned if the rancher faced a penalty. The eagle had survived electrical shock, sub-lethal lead exposure, emaciation, and a bill deformity, just to be toppled by a bullet.

Being shot or electrocuted are top threats to eagles. While small birds can zip between powerlines without brushing a feather, eagles more easily close the deadly gap between wires. If they touch two electrified pieces of powerline simultaneously, or they contact electrical equipment and a grounded object, they complete a very hot circuit. Even if they avoid contact, high voltage lines can arc through the air.

Then there are firearms. People have blasted away at raptors for as long as firearm companies have made guns. Despite protections from what are likely America's most well-known wildlife laws, lawbreakers kill eagles for sport. Others shoot eagles to sell the feathers for profit. On the black market, a single tail feather will fetch $100; an entire carcass will rake in $1,500. On Hawk Mountain, where Rob held his first Cooper's hawk, gunners once sat atop the mountain and killed hawks by the hundreds, spurred on by the 1929 incentive funded by the Pennsylvania Game Commission of offering a $5 bounty on northern goshawks to diminish predation on ruffed grouse.

An eagle's propensity to tussle with livestock, although uncommon, has motivated a barrage of gunfire from ranchers over the years. Bald eagles do occasionally attack lambs, but golden eagles probably sink their talons into livestock more frequently. One of the most extreme cases of livestock depredation came off a single ranch in New Mexico. Between 1987 and 1989, golden eagles killed twelve calves and injured sixty-one more, equaling a $20,000 loss. Similarly, sheep ranchers near Dillon, Montana, lost $38,000 worth of lambs in 1974 alone.

These accounts of large-scale predation events do make a case for why livestock owners may despise eagles. In 1951, Time magazine published a photo of a man riding in a plane, aiming down the barrel of a shotgun at an eagle. In the mid-twentieth century, gunning them down from the sky enabled a large-scale slaughter. Bald eagles were already protected but goldens weren't. Since the dark-headed juvenile balds could easily be mistaken for goldens, lawmakers incorporated goldens into the Bald Eagle Act in 1962. Shooting any eagle in the United States became one of the highest forms of wildlife crime. Today, people still insist on blasting eagles, but they risk facing massive penalties, such as a $100,000 fine and up to a year in prison.

Poisoning (Unrelated to Lead)

The penalties can be stiff for unintentionally killing eagles. In 2016, a North Dakota buffalo rancher spread Rozol, an anticoagulant rodenticide, to thin out the prairie dog population. Anticoagulant rodenticides disrupt the vitamin K cycle, a process responsible for helping blood clot. Excessive exposure causes a suite of nasty symptoms that stack up before death, including anemia, widespread bruising, hemorrhage, and bleeding from the nose, gums, gastrointestinal tract, and rectum. The rancher and his hired help weren't certified to spread this type of poison. They made a second mistake by scattering it aboveground instead of down burrows. The product's label says that users must check for dead animals and bury them to prevent secondary poisonings. The Rozol killed six bald eagles, and the rancher was required to pay $9,800 restitution per eagle; that's $58,800, on top of a $50,000 fine.

Rodenticides can crawl across the food web. When a rodent nibbles off bait, the animal's body may contain enough vitamin K to operate smoothly for a while. That gives the victim time to further drench its insides with poison. As the rodent begins hemorrhaging and becomes desperate, it might behave recklessly, such as by stumbling across a grassy field under a raptor's shadow. If the raptor eats the toxic prey and dies, the poison can jump to a third victim.

Raptor View doesn't test eagles for anticoagulant rodenticides, but a 2021 study found that 82% of the eagle livers collected from across the country contained some form of a rodenticide. Not far from the Wild Skies Raptor Center in the Blackfoot Valley, three bald eagles died near a lake from an unknown rodenticide, orphaning one eagle. The orphan fell from the nest and fractured a shoulder bone. Brooke healed and released the orphan, feeling relieved when the eagle appeared months later on game cameras set for the Bitterroot Valley Winter Eagle Project.

Lead Poisoning

In addition to lead exposure killing eagles straight-up, sub-lethal exposure might make it tougher for eagles to navigate the gauntlet of human-related threats they face. It's like how driving drunk makes you prone to accidents. Or, how being drunk makes you liable to root around for your roommate's leftovers instead of opening your own can of ravioli.

The research is mixed on whether lead elevates an eagle's odds of dying from other causes. One study collected dead eagles from across the lower forty-eight, finding alarming amounts of lead in eagle livers, yet those concentrations didn't differ based on the bird's causes of death.

Researchers studying golden eagles in Sweden found different results. Lead concentrations in eagle livers tended to be higher for birds that starved, collided with trains or cars, or experienced another form of trauma. The researchers also outfitted sixteen golden eagles with GPS transmitters to measure their movement rates. Eagles with higher lead levels tended to move less and fly lower.

If lead exposure does chisel away an eagle's ability to kill prey, the predator might crank up its scavenging behavior. Rob built evidence for this theory with his work on the Rocky Mountain Front. The golden eagles Raptor View captured that were feasting on carrion tested higher for lead than the eagles fooled by the pigeon lure. And if "leaded" eagles preferentially seek out dead animals, they may make riskier decisions, such as eating roadkill as cars whoosh past. They might skulk around lambing grounds as a shepherd takes aim with a rifle. They could investigate a trapper's wolf bait or gather rodents poisoned by anticoagulant rodenticides. They might gobble an elk liver that a hunter discarded in a grassland. And if that liver hides lead fragments, that meal might fuel a positive feedback loop of lead exposure, reinforcing the eagle's appetite for scavenging.

10. AN EAGLE'S VALUE
(in Tugrik)

Smoke leapt from wood barbecues as flames charred skewers of mutton. Musicians played string instruments, blending what sounded like traditional Chinese with the fast-paced rhythms I hear in war movies set in the Middle East. Locals paraded around the dirt running track. They wore traditional Kazakh outfits, sporting vests and jackets dyed in reds and blues, embroidered with white and gold stitching. Onlookers sat on the bleachers in trendy sunglasses that cut the July sun. In the grassy field inside the track, men wearing camouflaged military uniforms were mounted atop horses, playing tug-of-war with a sheep pelt. Vultures circled overhead, which if they caught a wandering itch, could land in China, Russia, or Kazakhstan by the next day. We stood on Central Asia's doorstep in Mongolia's far western Bayan Ölgii Province, and the Nadaam Festival had just begun.

Before leaving the States for the month-long vacation, Bridgette and I emailed a tour agency in the city of Ölgii, asking for a nine-day trip with a driver who could get us into the Altai Mountains to fish and hike. Our timing coincided with the year's biggest festival, where locals compete in archery, wrestling, and horse racing. But it's mainly a gathering where herders leave their pastures for a weekend to party. The mountains could wait two days.

We rounded the track to the south end with our guide, Berik. A hooded golden eagle was standing in the dirt. The white base on its tail feathers marked it as a one-year-old. The tips of those feathers were tattered and sparse like the teeth on a comb. Without having a perch to stand on, the ground had abused its plumage. The eagle was tethered to a saker falcon. The falcon's dark mustache and pale, spotted breast reminded me of its desert cousin back home—the prairie falcon. Sakers are endangered, although Mongolia hosts a thriving population of breeding pairs compared to other countries. With a bit of slack in the rope, the falcon jumped into flight. The rope went taut and the falcon jerked back, slamming to the earth with fluttering wings and a puff of dust. I cringed. A thin man in a blue striped shirt stood behind the birds.

"Do you want to hold eagle?" Berik asked with a heavy Kazakh accent, almost sounding Russian.

In a snap decision, I agreed—but immediately regretted supporting this type of business. If the eagle's threadbare feathers and the endangered falcon weren't clear enough warning signs, a biologist later told me that when two birds of prey are tied together, the larger bird sometimes kills the smaller one. Good thing the eagle was hooded.

Berik spoke to the bird owner in Kazakh and turned back to me. "Two thousand tugrik to hold eagle."

I reached in my pocket and handed the man a small wad of bills to cover Bridgette and me—about two dollars. The bird owner untied the eagle from the falcon, lashing the smaller bird to a post. He handed me a leather glove big enough to fit my head. I slid it on my hand and the man motioned for me to crouch to the eagle's feet. He coaxed the bird onto my arm by pulling the leather straps, called jesses, that were attached to its ankles. The eagle walked on. I rose with the giant raptor and stretched out my arm like a tree branch. I imagined the eagle's yellow feet walking up toward my shoulder and its talons slashing deep into my bicep. Then I wondered how many tourists rush to Ölgii's hospital each year with eagle wounds.

I felt uneasy about treating this young eagle like an attraction at Disney World. It was Mongolia's version of riding an elephant in Thailand or cuddling a koala in Australia. The whole thing seemed wrong. "That's good," I blurted, keeping my mouth shut afterward, not wanting to offend anyone or tarnish whatever experience Bridgette was about to have.

Bridgette adorned the glove and then lifted the eagle. A smile blossomed on her face.

"Move your arm," Berik said, twisting his wrist to demonstrate.

Bridgette took his advice. The golden eagle shifted its feet on the moving perch, spreading its dragon-like wings above her head to balance itself.

"Good for photo," said Berik.

And it was. Locals even stopped to watch. She was wielding the species that helped make their culture so famous.

***

The bond between golden eagles and the people of Central Asia has spanned millennia. About 500 miles to the southwest near Almaty, Kazakhstan, there's a petroglyph on a canyon wall dating back to 1200-1300 BC. Scratched in the rock is a human standing between his horse and a large prey animal. Two hounds are biting the prey's head while an eagle appears to be attacking its back. Nearby in Kyrgyzstan, another image dated to 600 AD shows a horse rider with an unmistakable eagle perched on his hand. Kazakhs cast that same silhouette today in Mongolia's Bayan Ölgii Province, ascending snowy hills on horseback with their eagles to hunt foxes, hares, and wolves.

Compared to North American falconry, where raptors are flown to kill small prey, such as rabbits and ducks, an eagle's willingness to take larger animals makes them a practical means to acquire pelts that can be worn or sold. Falconer and author Stephen Bodio traveled to western Mongolia, reporting in his book *Eagle Dreams* that the exchange rate for one fox pelt was a sheep. A skilled falconer could fill his pastures and his family's stomachs with a well-trained eagle, all while reducing the number of predators that might attack his lambs and other livestock.

Kazakh falconers prefer female eagles for their larger size, according to Bodio. Some eagles weigh upward of 15 pounds (Mongolia's eagles run heavy). Birds caught as adults are often the best fliers and known to be family-friendly. Falconers catch them using similar methods as biologists in North America. A falconer locates a free-flying bird they'd like to train, then they place a carcass on the ground near a catch-net; they tether another eagle close by. This entices the wild bird to swoop and drive off the competition, but the bird instead collides with the net. If the subsequent training goes well, a falconer could be hunting with the new eagle only a month later.

Young eagles acquired from the nest were riskier. They have a reputation as being bold, aggressive, and more likely to attack dogs, livestock, and their human trainers. During Bodio's travels, eagle hunters frequently recounted a story about one trained eagle that had found the courage to attack a snow leopard. The eagle lost that skirmish, and the falconer mourned its death. He was waiting for the mountains to thaw so he could "bury her in the sky." Another eagle hunter told Bodio a tragic story about one man's eagle who "veered off, too hungry, and killed his eight-year-old son."

Despite my enthusiasm for eagles, Bridgette and I didn't ask our tour organizer to introduce us to an eagle hunter. I felt uncomfortable treating the tradition as a slideshow. Mongolia's landscapes and people were fascinating enough. Before we left for Mongolia, I printed a stack of photos showing our family, house, and Montana's scenery. I threw in the image of me holding the golden eagle that Rob had caught.

***

The Monday after the Nadaam Festival, we met our driver, Miras. He was a short man who wore a blue Ferrari jacket and cotton sweatpants. Miras had soft eyes, an easy smile, spoke Kazakh first and Mongolian second. I knew as much Mongolian as he did English, which was fewer than the 100 essential words and phrases my phone app offered. We didn't know where Miras would take us besides into the mountains. But we had a good feeling about him. Sometimes that's all you need.

Miras steered his Toyota Landcruiser west over a rutted two-track, and we left urban Mongolia behind at the crest of a gravelly mountain pass. He hummed to music that matched the rhythm of a galloping horse, identical to the observation Bodio made a decade and a half earlier. Miras paused his humming to hear the song's neighing horse. The car rocked violently for hours as we ascended mountain passes and descended into new sweeping valleys. Water lapped against the car doors as we forded rivers, at one point passing a half-submerged Russian van that had been abandoned midstream.

Throughout our time in Mongolia, we marveled at how well the landscapes resembled Montana. Here, in the Bayan-Ölgii Province, broad valleys rose to windswept hills and plateaus like in southwest Montana. In the Gobi Desert, fawn-colored badlands were streaked with reds and purples like the ones near Terry, Montana, where my game cameras captured burrowing owls scavenging prairie dogs. The forested mountains of central Mongolia were indistinguishable from the forested mountains of central Montana. When we returned from Mongolia, I'd play a game with my family by showing them landscape photos and asking, "Montana or Mongolia?" They almost always mistook Mongolia for Montana.

One significant difference was Mongolia's lack of fences, with livestock outnumbering people by about twenty to one. Sheep, goats, and yaks roamed pastures that climbed to the horizon. Wherever we saw livestock, we saw gers (commonly called yurts in the U.S.) that, from a distance, appeared as if the earth had sprouted giant marshmallows. When we closed the distance, golden eagles were sometimes perched outside the dwellings, tethered to posts like dogs.

It turned out Miras was taking us to a ger camp in Altai Tavan Bogd National Park. To the north, a glacier near the Russian border fed a series of lakes and rivers that teemed with grayling. The land to the east was rocky and too parched to support livestock. A windswept ridge undulated to the west above patches of timber, separating Mongolia from China. I imagined golden eagles rising in thermals above that border. The door to the main ger faced south, framing a landscape pockmarked by kettle lakes and mounds of glacial till. Sheep and goats grazed the grassy landscape to stubble.

Inside, we sat with a nomadic family on short, wooden stools around a knee-high table. They'd placed cups of hot tea and a bowl of hard cheese on the purple tablecloth. Miras chatted with the mom in Kazakh. Two young boys and a girl, ranging in age from six to sixteen, sat on the floor across the table from my wife and me. They each leaned in and stared.

I slipped the collection of photos from my jacket pocket and held them out. The oldest boy snatched them away. He divvied them among his siblings, and each kid shuffled through the pictures, seeing our family, the landscapes of Montana, and finally, the photo of me holding a golden eagle. The middle son smiled proudly and held it out for everyone to see. Miras bent toward us with his phone. He showed us a picture of a man with a dark, weathered face, holding a golden eagle. Miras showed us enough photos for us to realize the man was his father.

The next day, he drove us to another nearby ger. The dwelling was positioned in front of a kettle lake where two golden eagles watched us from the shoreline. A man wearing a cream-colored sweater soon emerged from the ger. He had young eyes, but the deep creases around his mouth suggested his face knew shadeless summers and punishing winters.

Slung over his arm, he had a black robe with golden embroidery along the collar, cuffs, and lapel. He also held a red hat with a fur underside to cover the ears. It was an eagle hunting outfit. He lifted the garb and nodded for me to dress.

Thoughts of Disney World tumbled through my head again, questioning if this man was a legitimate eagle hunter. I shrugged on his hunting outfit anyway. The thick fabric was meant for enduring the coldest winters in Asia, not a mid-summer day, so beads of sweat began rolling down my back. The man slid a hood over one of the eagle's heads and pulled it atop his gloved hand. He held it close to his smile, adoring it like his child. I slid my hand into the glove as the falconer pulled his hand out. He wrapped the bird's jess around my fist to remove slack. The eagle's toes splayed across my hand as she settled in. Once the falconer removed the bird's hood, the eagle's eyes swept across my face and drew in the landscape from her elevated perch. Compared to the eagle Bridgette and I held at the Nadaam Festival, the feathers on this one-year-old bird looked full and crisp. The eagle was vibrant. Bridgette snapped the most exotic staged photo of my life before we swapped places.

Miras later asked me to offer the man seven-dollars-worth of tugrik. I felt comfortable paying. Earning money must have been tough in this remote pocket of Asia. A true eagle hunter could supplement their income by selling pelts from their hunts and introducing tourists to golden eagles. Maybe that exposure translates into tourists becoming advocates for the species. Seeing eagles close up proved to be the turning point for me. Eagles possess a tangible value that's only overshadowed by their symbolic worth to the human imagination. It's why eagles appear on bank notes, at least eight national flags, company emblems, and clothing logos. It's why eagles symbolized the soul in Egyptian hieroglyphics. It's why Native Americans carve eagles on totem poles, blow whistles made from eagle bone, and wear eagle feathers during ceremonies. Today, dead eagles in the U.S. are shipped to the National Eagle Repository in Colorado to be parted out and distributed to Native Americans. An eagle mascot even promotes gun safety for the National Rifle Association.

After Bridgette held the eagle, the man invited us inside his ger and had us sit on a red rug. He and our guide joined us in the dim space. His wife, slender-faced and beautiful, placed a board of cheese in the center of the rug. We each took a slice. It was soft like mozzarella with a hint of smoke. I looked around as I chewed. Three fox pelts hung by their noses from the lattice frame that gave their ger structure. The eagles outside were apparently more than photo props. Next to the fox hides, a rifle leaned against the rounded wall. It looked small in caliber, perhaps a .22 long rifle. The man probably killed the foxes with that gun after his bird attacked them. Yet, I doubted his birds tasted those bullets or any lead. They likely only swallowed meat from livestock killed by a blade. I wondered if he knew that lead bullets were poisoning golden eagles in my country. That eagles were crippled on Montana hillsides indistinguishable from the hillsides he knew in Mongolia. I wished I had the vocabulary to tell him.

11. HUNTERS as ALLIES

The conference hall was abuzz with hunters and anglers. Men and women, wearing plaid shirts and camouflage jackets, dipped in and out from booths where vendors were selling everything from binoculars to watercraft. In our case, we were selling an idea.

I traveled to Boise, Idaho, to help Chris Parish and Leland Brown with their non-lead booth at the Backcountry Hunters & Anglers Rendezvous. We each wore a black mechanic's shirt bearing the name of Chris's and Leland's new organization: The North American Non-lead Partnership. Behind us, we had two tables displaying an assortment of lead and non-lead options. Two ballistics gels exhibited the fragmentation rates of the different bullets. In front of us stood two falconers, one holding an American kestrel and the other a Harris's hawk. Chris recruited them to help.

As people walked by, they pulled phones from the back pockets of their jeans to take photos of the birds. Nobody could resist seeing a raptor's eye up close. When the bystanders slid the phones back into their pockets, we drew them into conversations about lead bullets.

"Where I work in Montana's Bitterroot Valley," I'd say, "about 95% of the golden eagles we catch have elevated levels of lead in their blood." That statistic was fresh from a manuscript I was coauthoring with Raptor View. Rob Domenech had approached me at a Christmas party, asking if I wanted to help publish his eagle data. After running the numbers, we realized Raptor View's winter study had documented the highest prevalence of "leaded" golden eagles ever recorded.

Before the rendezvous in Boise, I had practiced responses for every criticism I might hear. After seeing Chris disarm that mustached man two years prior, I was now armed with a diploma from an online people skills course. But I didn't need a single de-escalation technique. People stopped to say, "I switched to copper because I heard about what lead was doing to raptors," or, "Glad you're here." Most often, parents who had seen x-rays of deer shot with lead bullets said, "I'm not going to feed my kids lead."

The years of outreach and raising awareness seemed to be working, although we still met plenty of hunters who were unaware of the issue.

"Do you have any partition bullets?" One guy from Texas asked after seeing the bullet display over my shoulder.

I reached for a clear, hard resin that encapsulated a partition bullet after it had been fired into water. At least a hundred tiny particles of lead were entombed in the small display.

"Holy smokes!" His eyes widened as if he were a raptor himself. "I've got to show my dad."

We were winning over hunters, conversation by conversation.

At the event, I had plenty of time to chat with Leland and Chris about the North American Non-lead Partnership. "We're looking to engage agencies and groups who are thought leaders that hunters listen to," said Chris. Whether it's a local rod and gun club or a state fish and game agency, the Partnership helps them understand how wildlife may be exposed to lead. "We want to be a support mechanism for these groups," said Chris. He and Leland had already coordinated successful outreach efforts in their home states, so they were confident they could help groups shape their messaging and avoid common pitfalls.

The Partnership developed a clever logo. The orange, white, and blue color scheme subtly reminds me of the American flag. A mule deer buck juts from the left and an eagle soars from the right. The logo's center shows the bottom of a rifle cartridge. "That primer–that catalyst for igniting the powder–is the action that makes things happen," Chris said. Inside the primer is the silhouette of a man resting a hand on a young hunter's shoulder. "I thought it was a really cool concept to think that those who came before us, those mentors we still have with us, and the next generation are where we need to focus." The logo evokes a sense that tradition is being passed on.

The Partnership's mission is charging forward. They've gathered support from nearly fifty groups, including Backcountry Hunters & Anglers, the Midwest and Northeast Associations of Wildlife Agencies, a clothing company called First Lite, and, of course, MPG Ranch.

***

That autumn, a series of events reinforced my impression that momentum was ramping up. First, schools and hunting groups began asking if I offered presentations on lead poisoning. Then, a videographer at Montana Fish, Wildlife and Parks reached out. He said he wanted to produce a video on lead and copper bullets for their Outdoor Report series. Their 2-minute clips run on Montana's evening news and are posted to social media. The opportunity offered tremendous outreach potential.

The videographer met me at MPG Ranch on a bright October morning. I had lugged a folding table into a field and set two ballistics gels on top. He filmed me at my shooting bench, with my rifle resting on a backpack, speaking about how the lead poisoning topic had evolved past science to a social issue. Afterward, I settled my crosshairs on the first ballistics gel, trying to avoid blowing apart the slow-motion camera set beside my target.

In the film, "The Copper Option," the videographer granted me the supremely satisfying job title of "shooting sports researcher," before blending footage of rifle blasts and a golden eagle tearing into a gut pile, all while making a case for why hunters should consider non-lead. The video hit TVs across Montana and Facebook about a week before the hunting season. It quickly gained tens of thousands of views on social media. The comments sections became a battleground, with supporters citing scientific literature to beat back detractors.

Three weeks later, an unfortunate event involving a new hunter named Hannah Leonard offered an opportunity to further amplify our non-lead outreach.

Hannah grew up in Missoula without much interest in hunting. By college, though, the idea of acquiring her own animal protein motivated her to enroll in a hunter's education class. Hannah soaked up hunting know-how by inserting herself into the local hunting community. She joined the Backcountry Hunters & Angler's Collegiate Chapter at the University of Montana, where she watched me give a talk about the eagle work at MPG Ranch and non-lead alternatives.

Her first hunting season, Hannah took advantage of Backcountry Hunters & Angler's Hunting Mentorship Program under the tutelage of my friend, Hannah Nikonow (who would later find the lead-poisoned golden eagle I visited at Wild Skies.) The *Hannahs* chased elk tracks that weekend but returned to Missoula without venison.

The following year, Hannah Leonard acquired an unlikely mentor: her high school boyfriend's uncle. The uncle held a moose tag, so he and Hannah focused on locating moose in the mornings. In the afternoons, they searched for Hannah's first deer. They mostly tromped around private lands called Block Management Areas, where the state pays the landowners to allow hunter access. One particular BMA saw heavy hunting pressure, with ATVs and trucks growling along the roads, their occupants trying to scrounge a deer within shooting distance of their rig.

Hannah and her hunting mentor took a different approach. They explored each ridge by foot. As they crested a coulee back to the truck one afternoon, instead of seeing a deer on the nearest hillside, they saw a golden eagle. "I'm no wildlife biologist, but I knew something was wrong when I saw that raptor hobbling on the ground," Hannah remembers. "It wasn't flushing when we got near it. We figured it had flown into the power lines that ran across the BMA." The eagle tried to jump on weak legs with its wings draping over the brown grass. Hannah's hunting partner swept a parka over the eagle and wrapped it snug.

Hannah's neighbor had volunteered at Wild Skies, so Hannah knew to call Brooke Tanner. Jesse Varnado from Wild Skies met the hunters on Interstate-90 and raced the eagle back to their intensive care unit. Brooke sent a message to Hannah that night, saying the eagle tested high for lead.

Hannah felt sick to her stomach when Brooke relayed the news. Ever since she'd seen my presentation at the university, lead poisoning had been in the back of her mind.

The eagle died two days later.

"Having that eagle in my hands—alive—then Brooke calling me and telling me it had passed was such a bummer. I had a lead bullet chambered."

News of the eagle whirled around Raptor View and MPG Ranch like a dark cloud. My keyboard received the outpour of my frustration. I typed as if my words could resurrect the eagle or spare Brooke from a festering grief. I shucked my standard neutral messaging to communicate a narrative the public seldom heard. The Missoulian newspaper published the guest column the following week on November 20, 2019.

Lead poisoning kills Golden Eagle, hunters adapting

On November 10, Wild Skies Raptor Center in Potomac admitted an adult Golden Eagle with the classic symptoms of lead poisoning. The bird's talons were clenched, and its wings drooped to the floor. His blood hardly had enough red blood cells to carry oxygen to keep him alive. Brooke Tanner, one of the rehabilitators, analyzed the eagle's blood and found lead concentrations beyond what her instrument could measure. Despite attempts to save him, the eagle died 36 hours later.

About three weeks before this event, Montana Fish, Wildlife and Parks produced a 2-minute video addressing the issue of lead poisoning in scavengers. In the video, I discussed how lead bullets fragment after they hit an animal. If those pieces of lead are left in a gut pile or carcass, the remains might poison scavengers, like eagles. The solution is easy: hunt with copper bullets that rarely fragment and are much less toxic.

Once the video hit social media, an overwhelming number of supporters thanked the agency for raising awareness. Meanwhile, a small handful of critics turned the issue political or questioned the science. Those detracting opinions can create the illusion that hunters are unwilling to face the problem, and they don't reflect the hunting community I know.

I've given dozens of presentations about lead poisoning and discussed the issue with hundreds of hunters. I've met only a few people who oppose the scientific evidence, and as information spreads, their numbers dwindle. Many hunters are already on board and shoot copper bullets. The remainder are eager to know more. They're often shocked to learn that 9 out of 10 eagles caught near Missoula in the winter have been exposed to lead. Many of these hunters see lead poisoning as a conservation challenge that they can overcome, and some argue that the reputation of hunters depends on it.

The hunting tradition and industry run deep in Montana, and many voices advocate for hunting with copper bullets. With every magazine article, podcast, and conversation, non-lead alternatives are becoming more mainstream. The eagle that died at Wild Skies Raptor Center shows that the challenge is ongoing, but change is happening fast.

If you are a hunter who uses copper bullets, speak up. Tell your friends who hunt. Tell your friends who don't hunt. Make it known that hunters care about all species of wildlife, not just the ones that end up on dinner plates. After all, can you guess who found the poisoned eagle near a hunting access site and brought it to the rehabilitator? Hunters.

***

A year later, Hannah Leonard emailed me a photo of her first deer, a whitetail with four points on each antler. Her copper bullet dispatched the buck before it fled a step. Her boyfriend, who had hurled lead his entire life, needed no more convincing. He liked how Hannah's copper bullet seemed to destroy less meat than lead.

In Hannah's email, she said she was wrapping up her master's thesis in Resource Conservation. She asked if I knew of any job opportunities dealing specifically with non-lead outreach. I didn't.

A few weeks later, I received a call from Bryan Bedrosian, Rob Domenech's raptor colleague in Wyoming. Bryan and a former board member from the Teton Raptor Center, Wendy Dodson, were initiating a non-lead outreach campaign called Sporting Lead-Free. Bryan had run an outreach program a decade earlier in Jackson, where he and his colleagues had provided free non-lead ammunition to hunters. As a result, the average blood lead level in bald eagles dropped. Now, Bryan and Wendy hoped to reinvigorate a similar effort, starting statewide in Wyoming, home to one of the densest golden eagle populations in the country.

Compared to the North American Non-lead Partnership that brings together agencies and organizations, Sporting Lead-Free fosters a community of individuals. Membership is free, and they hold free demonstrations. For instance, after they shoot a ballistics gel, they radiograph it with their portable x-ray instrument, so spectators can see the tiniest bits of lead dust. The organization also aims to engage new hunters and young minds by incorporating the relevant science into hunting education classes and high school science curriculum.

Sporting Lead-Free intends to partner with retailers, both locally and nationally, to improve the access and visibility of lead-free ammunition and fishing tackle. Ammunition boxes often don't say whether the bullets contain lead. Sporting Lead-Free plans to introduce a universal symbol for lead-free products. Retailers can then display that symbol on their shelving or attach it to their products. As Sporting Lead-Free deepens its roots in Wyoming, Bryan hopes the model can branch out to other states.

As Bryan told me about the program on the phone, he mentioned that Sporting Lead-Free needed an outreach coordinator. Did I know of anyone?

Hannah Leonard thinks her eagle story helped land her the job. Within months of assuming her new role, she had already packed her outreach arsenal with talking points and communication strategies. "You just have to find the values that people care most about," Hannah said, "whether it's wildlife, their own health, or saving meat."

***

Another non-lead project that's budding in Wyoming is called Hunters for Eagle Conservation. The biologists running it, Dr. Vince Slabe and Ross Crandall, are studying whether wind operators could offset eagle fatalities by distributing non-lead ammunition to hunters.

If a wind farm kills a golden eagle, the operators might face one the whopping fines associated with the Bald and Golden Eagle Protection Act. Or instead, wind farm operators can preemptively apply for a "take" permit from the U.S. Fish and Wildlife Service. The permit allows operators to incidentally kill a certain number of golden eagles, but because the Fish and Wildlife Service has a "no net loss" policy for the species, wind operators must enact measures that help curtail or offset possible fatalities.

Wind operators have traditionally retrofitted power lines as an offset. The retrofit can increase the distance between wires, make the wires more visible, or move utility poles from problem areas. Statistical models help decide how much retrofitting and cost is necessary to "save" one eagle. The work must occur near a wind farm, but there are only so many power lines available to retrofit. Wind operators and the U.S. Fish and Wildlife Service have expressed the need for other economically viable offset strategies.

"If you're putting money into offset programs, here's an opportunity to not only potentially benefit eagle populations but also support a very important group of conservationists: the hunting community," said Vince Slabe, a research wildlife biologist for Conservation Science Global, Inc., a wildlife research non-profit. Vince spent a decade catching raptors with Rob Domenech on the Rocky Mountain Front before earning his Ph.D. studying lead poisoning in wildlife.

Hunters that hold specific tags in three hunt areas between Cheyenne and Casper, Wyoming, receive a flyer in the mail directing them to the project's website. After the hunter plugs in their unique code and agrees to participate in a post-hunting season survey, they are redirected to a website where they can order two boxes of free non-lead ammunition from Selway Armory, a Montana-based firearms dealer.

"The first season was a smashing success," said Ross Crandall, co-investigator on the study and Wyoming-based biologist. Crandall had worked alongside Bedrosian in Jackson to implement their non-lead distribution program. That past experience undoubtedly helped steer the new project to a similar success. "We could have kept going if the money didn't run out."

In the first season, the project supplied 434 hunters with two boxes of non-lead ammunition. "Secondary to our research, the program also serves as a vehicle to get non-lead into people's hands," said Ross. "If 5% of the hunters that participate in our project like the results of non-lead ammo and make the switch voluntarily, that's a good thing for wildlife."

After Vince and Ross complete the second year of their study, they will model how much non-lead ammunition must be distributed to offset an eagle fatality. If their offset model is economically feasible, Vince and Ross hope to expand their efforts to other states.

"Hunters were excited about the program," said Ross. "They were sending us trophy shots. They were also very excited to have so many eagles using the landscape."

12. RELEASE

A half-hour drive south from MPG Ranch takes you to Willow Creek, a gurgling stream that drains the Sapphire Mountains. Most afternoons, a local married couple named Stephen and Gail walk their Bernese mountain dog and black lab along the gravel road beside the creek. During a February stroll, their dogs whiffed a rotten smell. Their noses tossed through the air, leading them to a roadside embankment. Stephen and Gail peered down between a tangle of branches and saw a dead deer near the creek. A golden eagle lay beside the carcass with outstretched wings.

"I thought the eagle was dead," Stephen told me when I visited him and Gail. "Its wings weren't moving. It was listless." But then the eagle swiveled its head.

The next day, Stephen and Gail spotted the eagle on the opposite side of the carcass. Soon thereafter, a neighbor noticed the eagle climbing the embankment, threading its droopy wings between a fortress of stems. The man grabbed two branches lying in a ditch and used them like giant chopsticks to escort the eagle across the road.

Stephen and Gail concluded the eagle must be sick or injured, so they networked until they rounded up Brooke Tanner's phone number at Wild Skies. Brooke relayed the news to Kate Stone at MPG Ranch, who was trained in handling raptors and lived a short distance from Willow Creek. Brooke told Kate to look uphill. When sickness washes over golden eagles, or when they've suffered an injury, goldens often climb to a vantage point. Not only can they see threats from the high ground, but steep slopes also offer them launching pads for takeoff.

Kate met Stephen and Gail an hour later. Sure enough, the eagle had ascended a hill and stood at the base of a ponderosa pine. The bird flared its wings and raised its golden hackles at the sight of the incoming humans. "I really thought I could get the eagle with my hand net," Kate said. But just before she took a swipe, the eagle jumped and glided downhill like a grouse, sailing over an aspen grove and banking toward a roadside thicket.

For nearly an hour, Kate, Stephen, and Gail searched beside downed trees. They parted branches to peek into possible hiding places. A small flock of songbirds was flitting between the branches of a double-trunked ponderosa pine. "Chick-a-dee-dee-dee. Chick-a-dee-dee-dee."

Kate knew the birds' alarm calls could mean a predator lurked nearby. Kate gazed up to see the eagle perched thirty feet above. Its wings dangled between branches. "The eagle looked awkward,"

Kate said, "like it was in a place it didn't want to be." Kate had no choice but to leave the eagle. The bird was too high up and a bucket truck couldn't reach the location. "It's an unfortunate reality when trying to capture injured birds—we often can't catch them when they're in decent shape. They need to be down and doing really poorly before we can get them help."

The next day, Rob Domenech and his uncle met Stephen and Gail for a second attempt. The eagle had climbed back up the hill, so they located the bird immediately. Rob approached with a salmon dip net that extended to over fifteen feet long. But the extension did no good—the eagle saw Rob and launched. It coasted over sagebrush and past a log home. Rob clenched his dip net and sprinted after the bird, weaving himself between shrubs and tree trunks. "My lungs were burning," Rob remembers. He watched the eagle land on flat ground, where it began running with loping steps. Unable to gather air under its wings, the eagle ducked under a fallen log and hid. Rob rushed over and slid the dip net over the raptor's head. He juggled a falconry hood from his pocket and covered the eagle's eyes. The bird was so weak its neck sagged from the extra weight.

After inspection, Rob determined the eagle was an adult female, and Stephen and Gail nicknamed the eagle "Willow." Unfortunately, Willow's lead level was high enough to kill her, so Brooke administered chelation therapy soon after receiving her.

"Willow took a while to turn around," Brooke recalls. But the fact that Willow survived a potentially lethal dose of lead and *did* turn around made her a success story. After the chelator had stripped much of the lead from her blood over multiple treatments, Brooke and Jesse started Willow on a physical therapy regime. "It's like how a human stuck in a bed for too long needs to move around to get their joints working." The rehabbers laid Willow on her back, pushing and pulling her legs in a bicycle motion. They held Willow's chest as she regained her balance on perches. Next, Brooke and Jesse set Willow in the avian equivalent of a Jolly Jumper for babies. A harness with overhead straps supported the eagle, so she could stretch her legs just far enough to stand. "When eagles are going after prey, they stick out their legs really far. They need that range of motion."

Willow kept progressing, so Brooke advanced her to an outside enclosure about as big as a bedroom. "The goal is to have a bird fly from the ground to the perch that's about four-feet high," Brooke said. Willow started small, first climbing atop a stump no taller than a jackrabbit. Brooke set chunks of venison and quail on the gravel floor, forcing Willow to hop off her rest and improve her mobility.

After Willow mastered perching, Brooke advanced the eagle to the flight enclosure. There, Willow could pump her wings from perch to perch. Brooke encouraged these flights by giving Willow friendly pushes into the air. As the eagle's flight muscles strengthened, Brooke began creancing Willow. A creance is essentially a dog run that falconers use to train raptors. Brooke stretched a 100-foot length of paracord between tree trunks and tethered Willow with a swivel. Those short flights winded Willow at first, but Brooke slowly escalated the duration of the sessions, and Willow's flight endurance soared.

One April afternoon, Rob texted me. He and Brooke were driving from Wild Skies to MPG Ranch to release Willow. Brooke had just fed her nuggets of venison and quail, so she wouldn't need to hunt her first day back in the Bitterroot; her only task would be locating a roosting tree. Rob also banded Willow and strapped on a GPS transmitter. Everyone was antsy to see the unit lay Willow's tracks on the map. If she was a migratory bird, Willow might zoom north to recoup lost time. Other migratory golden eagles from the valley were already riding thermals halfway up British Columbia, if not vying for nesting ledges on Alaskan cliffs. If Willow were a resident, maybe she'd search for her old mate if he hadn't already courted a new female.

I met Rob and Brooke at the ranch. Stephen and Gail drove up from Willow Creek—they were as emotionally tied to this bird as anyone. The plan was to release Willow on a grassy hillside that overlooks the Bitterroot Valley. That way, the eagle's first sight would be a familiar landscape. The weather forecast for the next day called for snow, but we wore sunglasses and light pullovers. The bright day would hopefully create updrafts that Willow could ride.

Brooke was whirling with excitement. After seeing lead poisoning strip the vitality from so many eagles, she had helped trickle life back into a bird that was ready to fly again.

"What's it like releasing a raptor you've gotten to know?" I asked.

"Getting an adult golden eagle back out is one of the main reasons I do this. So, it's pretty awesome." Brooke said with a contagious smile. "I look forward to seeing what she does out there."

Jesse Varnado from Wild Skies unloaded an animal crate from Brooke's car. A blanket was draped over the top. He hauled the container to the release site. Rob stood beside the crate with his phone, ready to film Willow's first independent flight in months. Brooke knelt and swept off the blanket. She unhinged the wire door. Our eyes stared into the dark opening with mounting anticipation. Nothing happened. Brooke peered into the crate. "She's facing away," Brooke said, reaching in and pressing her hand on Willow's back.

Willow turned at Brooke's touch and hopped out with a flutter of her wings. Willow stood and absorbed her onlookers. She took in the valley's landmarks: the jagged peaks above Bass Creek. The Bitterroot River running between cottonwood forests like a snaking north-south compass. A layer of snow and ice had buried the valley floor the last time her eyes brought the terrain into focus. Now, ground squirrels were skittering across the damp soil, nibbling the emerging grasses under a warm sun.

Willow lifted her tail feathers and squirted out a white splash of excrement. We all laughed. With that ballast jettisoned, Willow launched, quickly gaining ground clearance over the falling slope. Her wings caught a northerly updraft and she hooked behind us toward a ridge. With powerful and fluid wingbeats, Willow's golden silhouette elevated above the horizon.

***

Willow must have recognized her neighborhood because she returned to Willow Creek the next day. She roamed the Sapphire Mountains, hunting the edges of forests and meadows. She flew over the ski runs at Lost Trail Pass into Idaho, spending six days in the timber above the Salmon River. Willow returned to Montana near Darby, soaring over the grassy foothills that rose steeply above riverside cottonwood galleries.

Over the next month, Willow began moving less, concentrating her activity to south-facing slopes, possibly diving on ground squirrels as they scurried to their burrows.

Almost two months after Willow's release, her transmitter sent a location from the bottom of a steep valley that a wildfire had scorched treeless. Points from her transmitter popped up over the next ten days, all from the same area. Rob gained permission from the landowner to investigate, hoping Willow had ripped off her transmitter. Instead, he found Willow lying next to a fallen log, her wings outstretched and flattening the green grass. She died with clenched talons.

Six weeks later, Rob and Brooke received the necropsy results. Willow had no fat stores. Her breasts had shrunken and her liver had atrophied. Her entire gastrointestinal tract was empty. The examiners said she died from chronic lead exposure.

The lead concentration in Willow's liver was high, but not extreme. Maybe Willow scavenged ground squirrels shot with lead bullets. Maybe her bones and tissues had been carrying that lead since before Brooke rehabilitated her. Either way, Willow's history of lead exposure had likely caused permanent neurological damage that crippled her ability to hunt.

Willow also tested positive for an anticoagulant rodenticide. The examiners couldn't determine how much that poison contributed to her death, but they suspected the concentration was too low to cause observable symptoms. Rodenticides often cause hemorrhaging, but Willow didn't show signs of suspicious bleeding.

The lab said that Willow's decreased movement was consistent with chronic disease and reduced energy reserves. As Willow slowly starved, lead may have mobilized from her bones back into her blood, further weakening her. When she spotted a poisoned rodent stumbling across a field, or one that had already bled out internally, Willow finally located a meal. That rat poison then blended with the toxic cocktail already churning through her veins, and Willow's success story ended in paralyzing agony.

EPILOGUE

The paw prints in the snow cradled five toes like a wolverine, but the roundness better resembled cat tracks. Further down the closed logging road, the prints forked into two paths. I crouched to the smaller set, removing a glove and running a finger over the crisp edge that separated the pad from toes—mountain lion. A cub must have been stepping in mama's tracks. Maybe two cubs.

The tracks veered off an embankment, and my boots plowed a fresh course down the road, now 1,500 vertical feet above my truck. If I climbed another fifteen minutes, I'd see the drainage where Rob Domenech netted Willow. Yet, I was hunting elk closer to where he found her lying dead in the grass.

The two-track contoured into a draw flush with willows before angling around a ridge of sparse timber. I had paused on ridges like these throughout the hunting season to watch golden eagles float on the updrafts above my route, their eyes scanning for prey with sudden glances. Peak migration had passed a month ago, once again filling the Bitterroot Valley with eagles from across western North America. Each time I saw an eagle's silhouette against the sky, I wondered how long it would take before I could marvel at eagles during hunting season without contemplating their lead burden.

As I hiked down the road, a young mule deer buck trailed four does through the trees. They stopped as I passed by, naively trusting a human wearing camouflage and blaze orange. I emerged from the forest onto a ridge that would steer me to my truck. But first, I walked another fifty yards to peek into the next draw. At least 100 elk were bedded on a bench speckled with stumps, with a small bull lying in the center. A yearling cow nearby raked its fuzzy head against a wild rose. The closest cows were grazing native bunchgrasses 350 yards away.

With the air current in my favor, I backpedaled and dropped down the far-side of the ridge to shave off 100 yards before hooking into a thicket of young firs. I reached an old road cut and squeezed through tree branches. When the limbs thinned and offered me a shooting window, I laid prone and rested my rifle across my backpack, a spare jacket crammed below my rifle's buttstock for added stability. Tucking the gun into my shoulder, I settled the crosshairs on a cow elk 260 yards below. A few exhales helped ease the tension in my shoulders, face, and legs, my body limp except for my finger tightening around the trigger.

The cow felt the blow to her ribs and the rest of the elk rose to their feet. As if being pushed down a funnel, the herd converged in the draw and pounded across a brushy slope.

Finding blood was tricky after four hundred hooves had churned the snow like mashed potatoes seasoned with dirt and willow leaves. Once I found the first sign she was hit, the trails began braiding through deadfall and shrubs. Between the red drips, I had trouble knowing which set of prints to follow. Tree sap had stained clumps of snow an amber that looked like diluted blood. I'd have to follow the trail methodically, otherwise, I might stray past a dead elk crumpled up in a thicket.

The blood trail brought me into a small basin. Movement behind firs grabbed my attention, so I lifted my binoculars. A female hunter with white hair was hiking uphill. She had covered her backpack and blaze orange vest with a sweeping white overcoat. Sensing my stare, she looked my way and spotted me waving.

I left the elk tracks and approached. Decked in white, the woman resembled a mythic sage plucked from a fantasy novel. Much of her gear appeared homemade and smelled musty from five feet away. Her rifle sling and binocular harness were braided from wool. She hiked with a lacquered walking stick. I asked if she saw the elk herd or crossed blood. She just missed them, she said, but figured my cow shouldn't be much farther. We wished each other good luck and parted ways. Retracing my steps, I heard an uphill whistle. The woman was pointing to red snow at her feet.

Back on the blood trail, I realized the herd had splintered into smaller bands. With a clearer path to follow, I cruised through the forest and found my elk. She fell on bare soil among century-old ponderosa pines and rock outcroppings that jutted like teeth from the earth.

The sun was dipping below the Bitterroot Mountains, so I slid on my headlamp. With game bags lying beside the elk, I shoved a canister of bear spray into my thigh pocket and hinged at the waist to begin the knife work.

An hour into the darkness, my blood-encrusted hands tied knots in the ends of four game bags loaded with lead-free meat. I fastened one bag to my backpack then hauled the others one by one to a pine seventy-five yards off, raising them a couple feet to cool. A bear's claws could easily slice the canvas open, spilling meat on the dirt. But without a buddy to help hoist, I knew I'd never heave it high enough. Instead, I peed on the tree trunk. For extra security, I surrounded the meat by weaving paracord between branches. It looked like a flimsy jungle gym that wouldn't defend against a stray wiener dog, so I mustered a second spritz of urine.

When I returned in the morning, a juvenile bald eagle coasted over the forest between me and the carcass. Another eagle hiding in the trees whistled a succession of sharp cries. I climbed above the rocky spires near the kill site to scope out what lay below. Predators hadn't breached my paracord fortress—the bloody game bags hung heavy. Several magpies perched atop the elk's ribcage, tearing niblets from cartilage. The birds squawked when they saw me navigating down the outcroppings.

The meat tree smelled like a gas station urinal, so I dropped my backpack clear of the splash zone. Bark flaked off branches as I lowered two game bags. After coiling the rope and loading meat for the first of two descents, I pulled a game camera from my pack. Near the carcass, I broke lower limbs off a pine and strapped the camera around its trunk.

After Kate Stone and Rob Domenech's success using game camera photos to build enthusiasm among landowners about scavengers, Kate and I began recruiting hunters to do the same. One hunter's camera caught a wolf pack gorging themselves with an elk carcass. Another hunter recorded a bobcat monopolizing an elk's innards. Eagles often landed, including golden eagles that Raptor View had tagged. The legacy of each successful hunt lasted far longer than bootprints in the snow or mud, and hunters—including me—enjoyed our spyglass into all the secret events that followed a kill.

By that evening, after I had stowed the last game bag in my truck bed, hung quarters off the pull-up bar in my garage, and began charring a tenderloin on the grill, a golden eagle's talons were gripping the elk's ribcage. Pink flesh stretched like bubble gum as the bird tugged tissue. The eagle shifted its stance and pulled bites from where my copper bullet pierced the elk. Three magpies eased in as the golden's crop ballooned. A second golden eagle landed twenty feet off and ran downhill to the elk's rear. As the two goldens chomped the carcass from both ends, a third golden landed near the elk's head. But without a safe path into the buffet, the eagle backed off without a bite.

Then snow fell overnight before melting at dawn under the yellow feet of eagles. An adult golden soon fanned its wings and nearly eclipsed the carcass as an immature golden touched down. The youngster inched closer until the older bird charged it, driving the competitor to the sky. Within five minutes, the young eagle dove with open talons and traded places with its elder. Goldens and balds sparred for the food cache until a snowstorm ushered in nightfall, once again burying meat and bones in powder. A fox slinked under parting clouds. The animal's black nose nudged snow away from the elk's belly before the fox's teeth excavated flakes of waxy fat.

In the twilight closing moments of the sixth day, a mountain lion pounced at the carcass and ousted a pair of golden eagles. Two spotted cubs scampered behind the mom. The mother began gnashing her teeth on the elk's sternum while a cub chewed on a vertebrate. The other cub walked across the pelvis and stuck its curious face into the ribcage, probing its blue eyes under the bone archways.

The evicted eagles flew above, their crops bulging with the last traces of red flesh.

Copper | Lead

FURTHER READING about
NON-LEAD AMMUNITION

When someone begins shooting non-lead ammunition, they frequently say, "I made the switch!" I'm glad they changed to less toxic bullets, but I'm not fond of the phrase. It builds an impression that shooting a new bullet is a big event. It's not. You just buy a box of non-lead ammunition, sight in your rifle, and go hunting.

Readers may wonder: If switching to non-lead ammunition is so simple, why did the author write this section? The simple answer is because hunters are a methodical bunch. Many of them have already sorted out which bullets they prefer, and those projectiles have filled freezers season after season. So, if a hunter switches ammunition, they want to be certain their bullets will fly true and be lethal.

Your bullet is the first piece of gear that contacts the animal, so proper bullet selection is paramount. But bullet selection becomes irrelevant if you are a crummy shooter. Shooting skills are neither inherited nor constant over your hunting career. Instead, they deteriorate unless you're burning powder at the range on a regular basis.

I recommend practicing your shooting technique from various positions and rests. Shoot short, shoot long. Do it under pressure with people watching. See how rapidly you can deliver a follow-up shot. Know how to use your hunting pack and winter jacket as rifle rests. Only when you've mastered your shooting technique can you truly begin to appreciate the performance of a bullet placed well at the clutch moment.

Availability

Many local business stock non-lead ammunition, and those who don't may be willing to order it for you. Shopping online offers the best availability if you live in a state that allows online ammunition purchases. Some online stores, such as Midway USA, provide a search filter for "lead-free" ammunition. Hunters who shoot common calibers like the longtime favorites .270 Win or .300 Win Mag might be pleasantly surprised by the number of options they encounter. Folks who shoot more obscure calibers, like the 7x57 Mauser or 6mm Rem, might be underwhelmed when they find few if any choices.

Based on the appearance of ammunition boxes, it's not always apparent if the bullets are non-lead. Even if the bullet jacket appears to be copper or gilding metal, its core may be lead. Watch out for the following words and phrases that would indicate a lead bullet: bonded, partition, full metal jacket, soft nose, or cup and core. Adding to the confusion, boxes of non-lead ammunition usually contain a warning about lead exposure. That refers to the lead styphnate in the primer, not necessarily lead in the bullet.

The list below will help you sort out what to buy. It isn't comprehensive, but it does include what you're likely to find for loaded non-lead ammunition. Some ammunition manufacturers load bullets produced by other companies. For instance, Remington loads Barnes TSX bullets for their HTP Copper ammunition. Numerous other manufacturers produce high-quality non-lead bullets that must be handloaded. If you aren't set up for that, you can hire it out with a company such as Safari Arms.

Centerfire

Barnes Bullets (TSX, TTSX, TAC-TX, LRX, [loaded ammunition called "VOR-TX"])

Black Hills (Barnes Varmint Grenade, Barnes TSX)

Browning (BXS)

Federal Premium (Trophy Copper, Barnes TSX, Power Shock Copper, TNT Green)

Hornady (CX, GMX, NTX, MonoFlex)

Norma (Ecostrike and Evostrike)

Nosler (E-Tip)

Remington (HTP Copper, Hog Hammer; both loaded with Barnes Bullets)

Weatherby (Barnes TSX, Barnes TTSX, Barnes X)

Winchester (Deer Season XP Copper Impact, E-Tip, Ballistic Silvertip Lead-Free)

Shotgun Slugs, Muzzleloader Bullets, and Round Balls

Barnes Bullets (Expander MZ, Spit-Fire MZ, Spit-Fire TMZ, Spit-Fire T-EZ)

California Republic Bullets (DeerDestroyer, ElkEliminator, CaribouCrusher, AntelopeAnnihilator [all round balls])

Cutting Edge Bullets (MZL Raptor, Maximus)

Federal Premium (Trophy Copper Muzzleloader, Trophy Copper Sabot Slug)

Remington (Premier Accutip, Premier Expander Slug, Premier Copper Solid Sabot Slug)

Winchester (Deer Season XP Sabot Slug Copper Impact)

Rimfire

CCI (Copper-22, TNT Green)

Hornady (NTX)

Norma (ECO)

Winchester (Hornady NTX, Varmint LF)

Cost

Ammunition prices fluctuate, but in general, a box of lead ammunition for big-game runs $20-70, compared to $30-70 for non-lead. If you shoot premium lead bullets, such as the partition or bonded constructions, you won't notice much of a price difference when switching to non-lead.

If a hunter shoots one box of ammunition per year (although

they likely shoot more for practice), they might spend an extra $10 if they switch to non-lead. Compared to optics, a rifle, clothing, tags, fuel, and for some hunters, a case of beer, $10 seems negligible, especially when it keeps lead out of your meat and prevents accidental poisonings of wildlife. For hunters who shoot more than one box per year, they can practice with more inexpensive ammunition at the range, where lead is arguably less available to wildlife. Before heading afield, the hunter can sight in their rifle with non-lead.

Picking a grain weight and understanding twist rate

Many hunters report equal or better penetration with lighter copper bullets than heavier lead bullets. This is a good thing because factory loaded copper bullets will often be lighter than lead options.

Copper is less dense than lead. So, a copper bullet must be physically longer to achieve the same weight as a lead bullet. When a bullet travels through a rifle barrel, the spiraling grooves (rifling) inside the bore spin the bullet, stabilizing the projectile for better accuracy. But for a rifle to stabilize a longer bullet, the bore's grooves must rotate at a rate fast enough to do the job. You can find your rifle's "twist rate" in the specifications of the gun, usually reported online. Twist rates are given as a ratio, such as 1:10, meaning the bullet will complete one revolution per ten inches of barrel. For bullets that require unusually fast twist rates, such as 1:8, bullet manufacturers usually say so on their website or in a bullet manufacturer's reloading manual.

Luckily, manufacturers load ammunition that has a good chance of working in a factory gun. Unless you shoot a custom gun or are reloading a heavier non-lead bullet than what is sold in loaded ammunition, you can probably ignore the twist rate of your gun and shoot factory loaded non-lead with great results.

Sighting in with non-lead

Whenever you switch ammunition, your bullet's point of impact on the target will usually change. This holds true whether switching from lead to lead, or from lead to copper. So make sure to readjust your riflescope zero to match any new bullet.

A good practice is to fire three-shot groups at 100 yards and adjust your riflescope between groups. Use a solid shooting rest if you can and pick a day when the wind is calm. If you just mounted a scope, you may need to start at twenty-five yards and work out to greater distances. Once you're driving bullets within a two-inch-wide circle of the bullseye, shoot another group to confirm your gun's accuracy. It's easy to burn through a box of ammunition getting a rifle zeroed. Rather than viewing it as an assault to your wallet and shoulder, consider it an opportunity to practice good shooting technique. After all, shot placement is more critical than bullet selection. You may decide to zero your gun for other distances. Regardless, remember to practice at the distances you expect to shoot while hunting.

If your barrel becomes too hot to grab while sighting in, consider taking a break. A hot barrel can shoot differently than a cool barrel. Personally, I'll fire a different rifle between my three-shot groups or dry fire my gun. Dry firing is safe for centerfire rifles and it helps you master your trigger pull, acquire targets, and it reveals kinks in your technique, such as flinching. Plus, it's free.

Terminal performance

Multiple studies have concluded copper bullets damage tissue as well as lead. Take a Barnes TSX, for example. When it impacts its target, the tip of the bullet peels back like a banana into four petals. The bullet spins like a drill bit with razor edges as it drives through the animal. Most copper bullets retain at least 99% of their weight, so there's maximum penetration without premature breakup.

In a 2013 study published in *Science of the Total Environment*, German researchers measured wound diameters in various hunted species. The sizes of wound channels didn't differ between animals shot with lead or copper bullets. The study found no evidence that fragments, lead or copper, increased the size of the central wound channel. The average diameter of entry holes were often less than half an inch. Exit holes were about 7–8 square inches and didn't differ based on whether they were created by a lead or copper bullet.

Scandinavian researchers measured penetration in a 2019 study published in *Ambio*. Copper bullets passed through moose as often as lead bullets. The researchers also found that bullet construction played no role in the distance that moose and other big game fled after being shot. When the researchers measured bullet expansion, copper bullets tended to be broader than lead bullets.

Fouling

Fouling is the buildup of lead, copper, or powder residues in the barrel's rifling. It can deteriorate accuracy. I've heard folks claim that solid copper bullets foul barrels faster than lead-core bullets. I've also heard they foul the same as lead. Most lead bullets have a copper jacket, so whether you shoot lead or non-lead, copper will probably contact the barrel.

But the difference in hardness between lead and copper comes into play. Lead bullets are softer and more malleable than copper, allowing air to more easily escape between the bullet and bore, reducing barrel pressure. Some copper bullets have grooves (a.k.a. cannelures or drive bands) cut around the lower half of the bullet to reduce barrel pressure and fouling. Regardless, clean your firearm regularly, and know that a dirty bore can result in deteriorating accuracy. Numerous online videos show proper cleaning techniques. Be aware that bore solvent best removes carbon buildup, and you should use a copper solvent and a nylon brush to remove copper buildup.

Should you worry about ballistic coefficients?

A few summers ago, I ran a non-lead outreach booth at the Youth Conservation and Education Expo at The Teller Wildlife Refuge in Corvallis, Montana. One parent approached the booth. "I shoot a Berger bullet with a BC of almost .700," he said. "Show me a copper bullet that can do that."

First, the BC, or the ballistic coefficient, is a number that describes how aerodynamic the bullet is, or how well it slices through the air. Put simply, does it push air like a Jeep or cut through it like a Corvette?

After the bullet leaves the barrel, air molecules exert friction on the projectile, slowing it down. The higher the BC, the better the bullet resists drag and retains its velocity. Maintaining bullet speed translates to a flatter trajectory with more kinetic energy hitting the target. Manufacturers improve a bullet's BC by tapering the projectile's base to create a boat tail. They can taper the nose and add a polymer tip, too, which also helps initiate expansion. Ultimately, the bullet becomes more streamlined and looks like a pointed missile.

Most hunting bullets have BCs between .200 and .500. The copper bullets I most often chamber have a BC around .400. So, did the parent shooting the bullet with a BC of .700 make a case against copper bullets? No way. Copper bullets are already considered premium ammunition and are plenty aerodynamic. I worry that this modern obsession with ballistic coefficients is fueled by a long-range lust that puts extreme accuracy ahead of terminal performance and the health of scavenging wildlife. If you plan to shoot less than 300 yards, I wouldn't stress about your bullet's BC. If your shots fly past that distance, especially past 400 yards, then yes, having a highly aerodynamic bullet is a must. The next section continues this discussion.

Shooting non-lead at long-range

The definition of long-range depends on who you ask. Let's say it's 400 yards or greater for our purposes. Arguments swirl around whether it's ethical to shoot animals at these distances. For readers interested in these ethical arguments, I recommend they read Boone and Crockett's position statement on long range shooting. Here, I'd like to set the ethics aside to focus on the external ballistics (flight) and wound ballistics (terminal performance) of non-lead at long-range.

Copper bullets have an advantage over lead bullets in that they can be turned on a lathe. That allows bullet makers to easily customize their projectiles into various aerodynamic designs, with boat tails and pointed tips, test them at the range, and perfect them with small modifications. As long as your rifle stabilizes the bullet (as discussed previously regarding twist rates), the bullet should fly fine.

In Bryan Litz's book, *Applied Ballistics for Long-Range Shooting*, he points out that measured BCs for copper bullets can be lower than what a ballistics software might predict. The discrepancy results from the extra air friction non-lead bullets experience from being longer and having drive bands that help prevent fouling. These aren't downsides, though. They are merely factors a long-range shooter must consider.

But just because you can spank a gong at 700 yards with copper doesn't mean your bullet will perform on an animal. Bullets require minimum velocities to expand. Obviously, the farther the bullet travels, the higher the odds its velocity will creep below that expansion threshold (this is partly why BC is so important to long-range shooters). It's tough to find minimum expansion velocities for individual bullets without contacting the manufacturer. In general, though, a copper bullet will probably expand if it's traveling 1,800 feet per second or faster.

I sometimes read articles where the author uses these minimum velocities for expansion as a cut-off for whether a bullet will work or not. That's too simplistic. Bullets expand to varying degrees depending on velocity. Even if a bullet does expand at 700 yards, it will be carrying drastically less energy and killing power than it did at half that distance. That applies to lead bullets, too.

Some copper bullets do expand at lower velocities, and by extension, at longer range. The Barnes's LRX (Long Range Hunting) is reported to expand out to 700 yards. Cutting Edge Bullets produce an Extended Range Raptor bullet that sheds six petals after impact on animals up to 600 yards away. I don't have experience with them, but a friend of MPG Ranch, where I work, machines a similar copper bullet. Not only do his bullets fly beautifully at extreme distances, hunters have used them to drop dozens of elk between 400-650 yards.

Ultimately, you can shoot non-lead at long-range when hunting, but like with any bullet, you need to understand its external ballistics. You need to know whether its design will offer killing power past 400 yards. Your gun needs to be extremely accurate, too. And you, as the operator of that firearm, should own a healthy sense of restraint informed by dozens of days at the range, where the conditions tested your knowledge of wind deflection, bullet drop with angled shots, and other advanced shooting concepts not covered here.

In other words, you'd better know what you're doing. Most hunters will have the best success by keeping their shots under 200 yards.

***

Non-lead for small game and varmints

In the mid-2000s, a couple of researchers estimated that shooters killed millions of prairie dogs annually. Factor in gophers, coyotes, marmots, and other "varmint" species that are shot and left in the field, and there's a smorgasbord of carcasses available to scavengers. If those animals were shot with lead bullets, scavengers will ingest some of that lead and possibly experience the debilitating aftereffects.

Some shooters jokingly say that all guns are varmint guns. I can't disagree. An elk rifle will certainly hammer prairie dogs. But in this section, I'm going to focus on the most popular small game rifles: the .17 HMR, .22 long rifle (LR), and .22 centerfires.

.17 HMR

It's easy to love the .17 HMR. Its bullets depart the muzzle almost as quickly as a bullet shot from a big game rifle. The resounding *CRACK* that accompanies each trigger pull could make a .22 LR whimper. And its projectile, roughly half the size of a pencil eraser, flies flatter than a .22 LR bullet. Plus, the shooter feels virtually no recoil. The company CCI produces a non-lead bullet for the .17 HMR called the TNT Green. The bullet is accurate, lethal, and costs the same as lead. I highly recommend it.

.22 LR

When it comes to caliber battles, the .17 HMR usually loses to the .22 LR. Ammunition for the .22 LR is often cheaper and comes in a range of options, with lead bullets leaving the muzzle anywhere from ~800-1450 feet per second. It's a favorite for squirrel hunters, plinkers, kids, and even doomsday preppers, due to its lightweight ammunition and ability to quietly thump small game.

Even though the .22 LR is the most popular rifle in the world, non-lead options are scant. The only cartridge I've found available for purchase is the CCI Copper-22. Weighing in at 21 grains, the bullet is just over half the weight of most lead options for that caliber. The projectile is powdered copper compressed into a bullet with a hollow point. With a muzzle velocity of 1,850 feet per second, it outpaces lead bullets and should pack a wallop.

I tested the precision of the Copper-22 at fifty yards with five different rifles. My average three-shot groups varied been 0.7 and 2.1 inches, depending on the rifle. Not bad.

Other researchers, however, have reported lackluster results for the Copper-22. While culling invasive rabbits in Australia, biologists noted the copper bullets wounded animals more often than one type of lead bullet. Although, their average shot distance was forty-nine yards, possibly at the outer limits of the Copper 22's effective range. The copper bullet also shot less precisely and didn't appear to expand.

The Copper-22 has other limitations. Being a lightweight projectile, the wind will blow it off course easier than a heavier lead bullet. A friend of mine compares it to a wiffle ball. It comes off the bat fast but succumbs to air resistance sooner than a baseball.

If you decide to head afield with the Copper-22, I recommend shooting at short distances, such as twenty-five yards or less. Up close, accuracy issues will be less pronounced and the bullet should carry more energy than many lead .22 LR bullets.

.22 centerfires

Sometimes when I chat with people about shooting small game, their eyebrows raise when I tell them a .22 centerfire, such as a .223 Rem or .22-250 Rem, shoots the same diameter bullet as a .22 LR. Based on the name alone, this shouldn't be a surprise (caliber equals the bullet diameter, in inches here). But I think the confusion comes from the difference in cartridge size and its application.

The cartridge for a .22 LR is about as long as a fingernail. A cartridge for a .22 centerfire is massive by comparison, being about two-thirds the length of a pinky finger. That added cartridge space means more powder, faster flight, and extra energy.

Non-lead rounds for .22 centerfires are excellent if the rifle stabilizes the bullet. The Barnes Varmint Grenade has a copper jacket that's filled with powdered copper and tin. It detonates after entry, vaporizing gophers and minimizing pelt damage on furbearers, such as coyotes and bobcats. The Barnes TSX does the opposite by holding together and driving deep.

But non-lead rounds for .22 centerfires have drawbacks unrelated to their performance. I rarely find more than one option at sporting goods stores. That option is usually the Barnes TSX, costing twice as much as the box of lead sitting beside it. If I were to shoot a deer with a .223 Rem, I'd trust that bullet more than any. But when it comes to shooting gophers and prairie dogs, using this bullet could turn into an expensive day.

Yet, it's that high volume of shooting that makes the use of non-lead so ecologically important. Bullets fired from a .22-250 Rem, for example, can travel at a blistering 4,000 feet per second. High-velocity bullets can pepper carcasses with over a hundred metal fragments. If a prairie dog shooter leaves fifty of those carcasses in the field, a single day of shooting could poison countless scavengers.

Bibliography

Prologue

Herring, G., Eagles-Smith, C. A., Bedrosian, B., Craighead, D., Domenech, R., Langner, H. W., ... & Wolstenholme, R. (2018). Critically assessing the utility of portable lead analyzers for wildlife conservation. Wildlife Society Bulletin, 42(2), 284-294.

Snyder, N. F., & Wiley, J. W. (1976). Sexual size dimorphism in hawks and owls of North America (No. 20). American Ornithologists' Union.

Chapter 2: Capture

Beckmann, J. P., & Berger, J. (2005). Pronghorn hypersensitiviy to avian scavengers following golden eagle predation. Western North American Naturalist, 65(1), 133–135.

Bodio, S. (2012). An eternity of eagles: the human history of the most fascinating bird in the world. The Lyons Press.

Clark, W. (1805). September 11, 1805. Journals of the Lewis and Clark Expedition. https://lewisandclarkjournals.unl.edu

Doyle, B. (2019). One long river of song. Little, Brown and Company.

Katzner, T. E., Kochert, M. N., Steenhof, K., McIntyre, C. L., Craig, E. H., & Miller, T. A. (2020). Golden eagle (Aquila chrysaetos). In In Birds of the World (version 2.0). Cornell Lab of Ornithology.

Langner, H. W., Domenech, R., Slabe, V. A., & Sullivan, S. P. (2015). Lead and mercury in fall migrant golden eagles from western North America. Archives of Environmental Contamination and Toxicology, 69(1), 54-61.

Potier, S., Mitkus, M., Bonadonna, F., Duriez, O., Isard, P. F., Dulaurent, T., ... & Kelber, A. (2017). Eye size, fovea, and foraging ecology in accipitriform raptors. Brain, Behavior and Evolution, 90(3), 232-242.

Smith, J. B., Walsh, D. P., Goldstein, E. J., Parsons, Z. D., Karsch, R. C., Stiver, J. R., ... & Jenks, J. A. (2014). Techniques for capturing bighorn sheep lambs. Wildlife Society Bulletin, 38(1), 165-174.

Chapter 3: Before Unleaded Gas

Butler, L. J., Scammell, M. K., & Benson, E. B. (2016). The Flint, Michigan, water crisis: A case study in regulatory failure and environmental injustice. Environmental Justice, 9(4), 93-97.

Childhood Lead Poisoning Prevention. (2021). Centers for Disease Control and Prevention.

Ferraro, J. (1996). Sesame Street: Lead away!

Hong, S., Candelone, J. P., Patterson, C. C., & Boutron, C. F. (1994). Greenland ice evidence of hemispheric lead pollution two millennia ago by Greek and Roman civilizations. Science, 265(5180), 1841-1843.

Mason, L. H., Harp, J. P., & Han, D. Y. (2014). Pb neurotoxicity: neuropsychological effects of lead toxicity. BioMed research international, 2014.

Patterson, C., Ericson, J., Manea-Krichten, M., & Shirahata, H. (1991). Natural skeletal levels of lead in Homo sapiens sapiens uncontaminated by technological lead. Science of the Total Environment, 107, 205-236.

Retief, F. P., & Cilliers, L. (2006). Lead poisoning in ancient Rome. Acta Theologica, 26(2), 147-164.

Taylor, M. P., Forbes, M. K., Opeskin, B., Parr, N., & Lanphear, B. P. (2016). The relationship between atmospheric lead emissions and aggressive crime: an ecological study. Environmental Health, 15(1), 1-10.

Tong, S., Schirnding, Y. E. V., & Prapamontol, T. (2000). Environmental lead exposure: a public health problem of global dimensions. Bulletin of the world health organization, 78, 1068-1077.

Chapter 4: Lead Shot, Dead Ducks, and Poisoned Eagles to the Rescue

America's bald eagle population continues to soar. (2021). U.S. Fish and Wildlife Service. https://www.fws.gov.

Anderson, W. L., Havera, S. P., & Zercher, B. W. (2000). Ingestion of lead and nontoxic shotgun pellets by ducks in the Mississippi flyway. The Journal of Wildlife Management, 848-857.

Bellrose, F. C. (1959). Lead poisoning as a mortality factor in waterfowl populations. Illinois Natural History Survey Bulletin; v. 027, no. 03.

Friend, M., Franson, J. C., & Anderson, W. L. (2009). Biological and societal dimensions of lead poisoning in birds in the USA. Ingestion of lead from spent ammunition: implications for wildlife and humans. The Peregrine Fund, Boise, Idaho, USA.

Grinnell, G. B. (1894). Lead-poisoning. Forest and Stream, 42(6),117–118.

McTee, M. R., Mummey, D. L., Ramsey, P. W., & Hinman, N. W. (2016). Extreme soil acidity from biodegradable trap and skeet targets increases severity of pollution at shooting ranges. Science of the Total Environment, 539, 546–550.

Moore, J. N., & Luoma, S. N. (1990). Hazardous wastes from large-scale metal extraction. A case study. Environmental science & technology, 24(9), 1278-1285.

Mulhern, B. M., Reichel, W. L., Locke, L. N., Lamont, T. G., Belisle, A., Cromartie, E., Bagley, G. E., & Prouty, R. M. (1970). Organochlorine residues and autopsy data From bald eagles 1966-68. Pesticides Monitoring Journal, 4(140), 141–144.

Chapter 5: Bring Me a Carcass, Hold the Poison

Alagona, P. S. (2004). Biography of a" feathered pig": The California condor conservation controversy. Journal of the History of Biology, 37(3), 557-583.

Beasley, D. E., Koltz, A. M., Lambert, J. E., Fierer, N., & Dunn, R. R. (2015). The evolution of stomach acidity and its relevance to the human microbiome.

PloS one, 10(7), e0134116.

Bedrosian, B., Craighead, D., & Crandall, R. (2012). Lead exposure in bald eagles from big game hunting, the continental implications and successful mitigation efforts. PLoS One, 7(12), e51978.

Cruz-Martinez, L., Redig, P. T., & Deen, J. (2012). Lead from spent ammunition: a source of exposure and poisoning in bald eagles. Human-Wildlife Interactions, 6(1), 94-104.

Domenech, R., Shreading, A., Ramsey, P., & McTee, M. (2020). Widespread Lead Exposure in Golden Eagles Captured in Montana. The Journal of Wildlife Management.

Fachehoun, R. C., Levesque, B., Dumas, P., St-Louis, A., Dube, M., & Ayotte, P. (2015). Lead exposure through consumption of big game meat in Quebec, Canada: risk assessment and perception. Food Additives & Contaminants: Part A, 32(9), 1501-1511.

Finkelstein, M., Kuspa, Z., Snyder, N. F., & Schmitt, N. J. (2020). California Condor (Gymnogyps californianus). In Birds of the World (version 1.). Cornell Lab of Ornithology.

Flores, D. (2016). Coyote America. Basic Books.

Ford, S. (2010). Raptor gastroenterology. Journal of Exotic Pet Medicine, 19(2), 140-150.

Haig, S. M., D'Elia, J., Eagles-Smith, C., Fair, J. M., Gervais, J., Herring, G., ... & Schulz, J. H. (2014). The persistent problem of lead poisoning in birds from ammunition and fishing tackle. The Condor: Ornithological Applications, 116(3), 408-428.

Hunt, W. G., Burnham, W., Parish, C. N., Burnham, K. K., Mutch, B. R. I. A. N., & Oaks, J. L. (2006). Bullet fragments in deer remains: implications for lead exposure in avian scavengers. Wildlife Society Bulletin, 34(1), 167-170.

Iqbal, S., Blumenthal, W., Kennedy, C., Yip, F. Y., Pickard, S., Flanders, W. D., ... & Brown, M. J. (2009). Hunting with lead: association between blood lead levels and wild game consumption. Environmental Research, 109(8), 952-959.

Katzner, T. E., Kochert, M. N., Steenhof, K., McIntyre, C. L., Craig, E. H., & Miller, T. A. (2020). Golden eagle (Aquila chrysaetos). In In Birds of the World (version 2.0). Cornell Lab of Ornithology.

Langner, H. W., Domenech, R., Slabe, V. A., & Sullivan, S. P. (2015). Lead and Mercury in Fall Migrant Golden Eagles from Western North America. Archives of Environmental Contamination and Toxicology, 69(1), 54–61.

Macaulay, L. (2016). The role of wildlife-associated recreation in private land use and conservation: Providing the missing baseline. Land Use Policy, 58, 218-233.

Markandya, A., Taylor, T., Longo, A., Murty, M. N., Murty, S., & Dhavala, K. (2008). Counting the cost of vulture decline—an appraisal of the human health and other benefits of vultures in India. Ecological economics, 67(2), 194-204.

Martin, D. (1996). On the cultural ecology of sky burial on the Himalayan Plateau. East and West, 46(3/4), 353-370.

Oaks, J. L., Gilbert, M., Virani, M. Z., Watson, R. T., Meteyer, C. U., Rideout, B. A., ... & Khan, A. A. (2004). Diclofenac residues as the cause of vulture population decline in Pakistan. Nature, 427(6975), 630-633.

Ogada, D., Shaw, P., Beyers, R. L., Buij, R., Murn, C., Thiollay, J. M., ... & Sinclair, A. R. (2016). Another continental vulture crisis: Africa's vultures collapsing toward extinction. Conservation Letters, 9(2), 89-97.

Oro, D., Genovart, M., Tavecchia, G., Fowler, M. S., & Martínez-Abraín, A. (2013). Ecological and evolutionary implications of food subsidies from humans. Ecology Letters, 16(12), 1501–1514.

Rogers, T. A. (2010). Lead exposure in large carnivores in the greater Yellowstone ecosystem. University of Montana.

Stauber, E., Finch, N., Talcott, P. A., & Gay, J. M. (2010). Lead poisoning of bald (Haliaeetus leucocephalus) and golden (Aquila chrysaetos) eagles in the US inland Pacific Northwest region—An 18-year retrospective study: 1991–2008. Journal of Avian Medicine and Surgery, 24(4), 279-287.

U.S. Fish and Wildlife Service and U.S. Census Bureau. (2018). 2016 National Survey of Fishing, Hunting, and Wildlife-Associated Recreation.

Wayland, M., & Bollinger, T. (1999). Lead exposure and poisoning in bald eagles and golden eagles in the Canadian prairie provinces. Environmental Pollution, 104(3), 341-350.

Chapter 6: The Copper Option

Carr, J. (2018). The terminal list: a thriller. Atria/Emily Bestler Books

Christian Franson, J., Lahner, L. L., Meteyer, C. U., & Rattner, B. A. (2012). Copper Pellets Simulating Oral Exposure to Copper Ammunition: Absence of Toxicity in American Kestrels (Falco sparverius). Archives of Environmental Contamination and Toxicology, 62(1), 145–153.

Massaro, P. (2015). Understanding ballistics: complete guide to bullet selection. Gun Digest Books.

McCann, B. E., Whitworth, W., & Newman, R. A. (2016). Efficacy of non-lead ammunition for culling elk at Theodore Roosevelt National Park. Human–Wildlife Interactions, 10(2), 268–282.

Rees, C. (2010). Triple-Shock X-Bullet. Rifleshooter.

Seng, P. T. (2006). Non-lead ammunition program hunter survey. Final Report to the Arizona Game and Fish Department.

Chapter 7: Small Game, Big Exposure

Ellis, D. H. (2013). Enter the realm of the golden eagle (N. Miller (Ed.)). Hancock House Publishers.

Frey, D. (2017). WSB: Seeking bullets lethal to small mammals, not scavengers. The Wildlife Society.

Knopper, L. D., Mineau, P., Scheuhammer, A. M., Bond, D. E., & McKinnon, D. T. (2006). Carcasses of shot Richardson's ground squirrels may pose lead hazards to scavenging hawks. Journal of Wildlife Management, 70(1), 295–299.

McTee, M., Hiller, B., & Ramsey, P. (2019). Free lunch, may contain lead: scavenging shot small mammals. The Journal of Wildlife Management, 83(6)

McTee, M., Young, M., Umansky, A., & Ramsey, P. (2017). Better bullets to shoot small mammals without poisoning scavengers. Wildlife Society Bulletin, 41(4), 736–742.

Pauli, J. N., & Buskirk, S. W. (2007). Recreational shooting of prairie dogs: a portal for lead entering wildlife food chains. Journal of Wildlife Management, 71(1), 103–108.

Reeve, A. F., & Vosburgh, T. C. (2005). Recreational shooting of prairie dogs. In J. L. Hoogland (Ed.), Conservation of the black-tailed prairie dog (pp. 139–156). Island Press.

White, C. (2005). Hunters ring dinner bell for ravens: experimental evidence of a unique foraging strategy. Ecology, 86(4), 1057–1060.

Chapter 8: Legislate or Educate?

Blake, J. (2017). USFWS issues Director's Order for use of nontoxic ammunition. The Wildlife Society.

Booms, T. L., Paprocki, N. A., Eisaguirre, J. M., Barger, C. P., Lewis, S. B., & Breed, G. A. (2021). Golden Eagle Abundance in Alaska: Migration Counts and Movement Data Generate a Conservative Population Estimate. Journal of Raptor Research, 55(4), 496-509.

California State Assembly. (2018). Ridley-Tree Condor Preservation Act. In Assembly Bill No. 821.

Center for Biological Diversity, American Bird Conservancy, Association of Avian Veterinarians, Project Gutpile, & Public Employees for Environmental Responsibility. (2010). Petition to the Environmental Protection Agency to ban lead shot, bullets, and fishing sinkers under the Toxic Substances and Control Act.

Center for Biological Diversity, Cornell Laboratory of Ornithology, Project Gutpile, Loon Preservation Committee, Loon Preservation Committee, The Trumpeter Swan Society,..., & Zumbro Valley Audubon Society. (2012). Petition to the Environmental Protection Agency to regulate lead bullets and shot under the Toxic Substances Control Act.

Claims and truths. (2014). Huntfortruth.org.

Epps, C. W. (2014). Considering the switch: Challenges of transitioning to non-lead hunting ammunition. The Condor, 116(3), 429–434.

Finkelstein, M. E., Doak, D. F., George, D., Burnett, J., Brandt, J., Church, M., ... & Smith, D. R. (2012). Lead poisoning and the deceptive recovery of the critically endangered California condor. Proceedings of the National Academy of Sciences, 109(28), 11449-11454.

Katzner, T. E., Kochert, M. N., Steenhof, K., McIntyre, C. L., Craig, E. H., & Miller, T. A. (2020). Golden eagle (Aquila chrysaetos). In In Birds of the World (version 2.0). Cornell Lab of Ornithology.

Katzner, T., Smith, B. W., Miller, T. A., Brandes, D., Cooper, J., Lanzone, M., ... & Bildstein, K. L. (2012). Status, biology, and conservation priorities for North America's eastern Golden Eagle (Aquila chrysaetos) population. The Auk, 129(1), 168-176.

Kelly, T. R., Bloom, P. H., Torres, S. G., Hernandez, Y. Z., Poppenga, R. H., Boyce, W. M., & Johnson, C. K. (2011). Impact of the California Lead Ammunition Ban on Reducing Lead Exposure in Golden Eagles and Turkey Vultures. PLoS ONE, 6(4), e17656.

Kochert, M. N., Steenhof, K., Carpenter, L. B., & Marzluff, J. M. (1999). Effects of fire on golden eagle territory occupancy and reproductive success. The Journal of Wildlife Management, 773-780.

RMEF joins fight against threat to use of traditional ammunition. (2014). Rocky Mountain Elk Foundation.

Smith, J. P., Farmer, C. J., Hoffman, S. W., Kaltenecker, G. S., Woodruff, K. Z., & Sherrington, P. F. (2008). Trends in autumn counts of migratory raptors in western North America. State of North America's birds of prey. Series in Ornithology, 3, 217-251.

The Humane Society of the United States, The Fund for Animals, Defenders of Wildlife, Natural Resources Defense Council, Wildlife Conservation Society, The International Wildlife Rehabilitators Association,...& Warren, A. (2014). Petition for rulemaking to require the use of nontoxic ammunition.

US Fish and Wildlife Service. (2016). Bald and golden eagles: population demographics and estimation of sustainable take in the United States, 2016 update. Division of Migratory Bird Management, Washington, DC USA.

Chapter 9: American Eagle

Avery, M. L., & Cummings, J. L. (2004). Livestock depredations by black vultures and golden eagles. Sheep and Goat Research Journal, 19, 58–63.

Bodio, S. (2012). An eternity of eagles: the human history of the most fascinating bird in the world. The Lyons Press.

Ecke, F., Singh, N. J., Arnemo, J. M., Bignert, A., Helander, B., Berglund, Å. M., ... & Hörnfeldt, B. (2017). Sublethal lead exposure alters movement behavior in free-ranging golden eagles. Environmental Science & Technology, 51(10), 5729-5736.

Franson, J. C., & Russell, R. E. (2014). Lead and eagles: demographic and pathological characteristics of poisoning, and exposure levels associated with other causes of mortality. Ecotoxicology, 23(9), 1722-1731.

Hall, J. (2017). Eagles are being killed for black market body parts. National Geographic.

Herring, G., Eagles-Smith, C. A., & Buck, J. (2017). Characterizing golden eagle risk to lead and anticoagulant rodenticide exposure: a review. Journal of Raptor Research, 51(3), 273-292.

Kagan, R. A. (2016). Electrocution of raptors on power lines: a review of necropsy methods and findings. Veterinary Pathology, 53(5), 1030-1036.

Katzner, T. E., Kochert, M. N., Steenhof, K., McIntyre, C. L., Craig, E. H., & Miller, T. A. (2020). Golden eagle (Aquila chrysaetos). In In Birds of the World (version 2.0). Cornell Lab of Ornithology.

Langner, H. W., Domenech, R., Slabe, V. A., & Sullivan, S. P. (2015). Lead and mercury in fall migrant golden eagles from western North America. Archives of Environmental Contamination and Toxicology, 69(1), 54-61.

Loss, S. R., Will, T., & Marra, P. P. (2013). Estimates of bird collision mortality at wind facilities in the contiguous United States. Biological Conservation, 168, 201-209.

Marr, N. V., Edge, W. D., Anthony, R. G., & Valburg, R. (1995). Sheep carcass availability and use by bald eagles. The Wilson Bulletin, 251-257.

Midwest Region-bald and golden eagles. (2019). U.S. Fish and Wildlife Service.

ND buffalo rancher admits killing 6 SD bald eagles with prairie dog poison. (2020). Capital Journal.

Niedringhaus, K. D., Nemeth, N. M., Gibbs, S., Zimmerman, J., Shender, L., Slankard, K., ... & Ruder, M. G. (2021). Anticoagulant rodenticide exposure and toxicosis in bald eagles (Haliaeetus leucocephalus) and golden eagles (Aquila chrysaetos) in the United States. Plos one, 16(4), e0246134.

Niemeyer, C. (2010). Wolfer. Bottlefly Press.

Phillips, R. L., & Blom, F. S. (1988). Distribution and magnitude of eagle/livestock conflicts in the western United States. Proceedings of the Thirteenth Vertebrate Pest Conference.

Smallwood, K. S. (2013). Comparing bird and bat fatality-rate estimates among North American wind-energy projects. Wildlife Society Bulletin, 37(1), 19-33.

US Fish and Wildlife Service. (2016). Bald and golden eagles: population demographics and estimation of sustainable take in the United States, 2016 update. Division of Migratory Bird Management, Washington, DC USA.

Watson, J. W., Vekasy, M. S., Nelson, J. D., & Orr, M. R. (2019). Eagle visitation rates to carrion in a winter scavenging guild. The Journal of Wildlife Management, 83(8), 1735-1743.

Watt, B. E., Proudfoot, A. T., Bradberry, S. M., & Vale, J. A. (2005). Anticoagulant rodenticides. Toxicological reviews, 24(4), 259-269.

Chapter 10: An Eagle's Value (In Tugrik)

Baljmaa, T. (2019). Mongolia has 70.9 million livestock animals counted. Montsame.

Bodio, S. (2003). Eagle dreams. The Lyons Press.

Nalewicki, J. (2016). Inside a remarkable repository that supplies eagle parts to Native Americans and science. Smithsonian Magazine.

Orta, J., Boesman, P. F. D., Sharpe, C. J., & Marks, J. S. (2020). Saker Falcon (Falco cherrug). In Birds of the World (version 1.). Cornell Lab of Ornithology.

Chapter 11: Hunters as Allies

Allison, T. D., Cochrane, J. F., Lonsdorf, E., & Sanders-Reed, C. (2017). A review of options for mitigating take of Golden Eagles at wind energy facilities. Journal of Raptor Research, 51(3), 319-333.

Bortolotti, G. R. (1984). Trap and poison mortality of golden and bald eagles. The Journal of Wildlife Management1, 48(4), 1173–1179.

Domenech, R., Shreading, A., Ramsey, P., & McTee, M. (2020). Widespread Lead Exposure in Golden Eagles Captured in Montana. The Journal of Wildlife Management.

McTee, M. (2019). Lead poisoning kills golden eagle, huntings adapting. Missoulian.

Further Reading about Non-lead Ammunition

Hampton, J. O., DeNicola, A. J., & Forsyth, D. M. (2020). Assessment of Lead-Free. 22 LR Bullets for Shooting European Rabbits. Wildlife Society Bulletin, 44(4), 760-765.

Litz, B. (2015). Applied ballistics for long range shooting. Applied Ballistics.

McTee, M., & Ramsey, P. (2022). Is lead-free. 22 long rifle ammunition worth a shot?. Wildlife Society Bulletin, e1255.

Stokke, S., Arnemo, J. M., & Brainerd, S. (2019). Unleaded hunting: Are copper bullets and lead-based bullets equally effective for killing big game? Ambio, 48(9), 1044-1055.

Trinogga, A., Fritsch, G., Hofer, H., & Krone, O. (2013). Are lead-free hunting rifle bullets as effective at killing wildlife as conventional lead bullets? A comparison based on wound size and morphology. Science of the Total Environment, 443, 226-232.

Acknowledgments

First, I owe a big thanks to MPG Ranch and Dr. Philip Ramsey for making this book possible. For a decade, they have supported lead-oriented research and helped fund outreach efforts. I greatly appreciate Daniel Rice for seeing the merit in this story and shepherding it through the publishing process.

Wilted Wings has always felt like a collaborative project. The book started as an essay in Phil Condon's Environmental Writing class at the University of Montana. My classmates offered outstanding feedback before I distributed the piece among my writer friends. "Oh my gosh, he's writing a book," Hannah Nikonow said after reading the essay. Although I had no intention of actually writing a book, the idea had been tossing around my mind as a pipedream. So, thanks Hannah for nudging the pipedream toward reality. Regarding the essay, I'd like to thank Joe Ballerini and Ryan Sparks at *Strung* magazine for helping it reach a broad audience.

An enormous thanks goes to Zack Williams, Brad Lane, Paul Queneau, and Noah Davis, who monthly, helped straighten the trajectory of each chapter while elk burgers sizzled on the grill. Noah deserves an extra high-five for immersing himself in the complete manuscript, identifying gaps in the narrative, and coming up with the book's title.

My honorary family at MPG Ranch offered encouragement, guidance, and edits throughout the writing of the book. Early versions greatly benefited from feedback offered by William Blake,

Emily Martin, Dr. Dan Mummey, Marirose Kuhlman, Craig Jourdonnais, Lorinda Bullington, and Eric "Kerr" Rasmussen. Beau
Larkin, Dr. Ylva Lekberg, Kate Stone, and Dr. Philip Ramsey not
only read early chapters, but they also provided feedback on the
complete manuscript. I couldn't imagine being steeped among a
more friendly and brainy group of naturalists.

Rob Domenech and Adam Shreading at Raptor View Research
Institute deserve big hugs for sharing their stories and data. Seeing a raptor up close can change a person's life. And the team
at Raptor View, including Brian Busby, Bracken Brown, Bridget
Creel, Beth Mendelsohn, Sarah Norton, Mary Scofield, Danny
Stark, Jack Toriello, and Tyler Veto have introduced hundreds of
students, children, and adults (including me) to eagles and other
raptors at MPG Ranch. I want to thank these biologists for their
exceptional raptor wrangling, their utmost care of the birds, and
for letting me hang around for when they blast the net gun.

I'm grateful to Brooke Tanner and Jesse Varnado for answering
my never-ending list of questions and allowing me to visit Wild
Skies Raptor Center. I extend the same thanks to Becky Kean
and Jordan Spyke at the Montana Raptor Conservation Center.
Rehabilitating sick and injured raptors is emotional work, so I appreciate your efforts to keep these inspiring birds aloft.

Numerous biologists, ballistics experts, raptor folk, and other
helpful acquaintances were generous with their time, offering me
interviews, their perspectives, photos, outreach opportunities, and
sometimes edits. These people include: Estelle Shuttleworth, Ross
Crandall, Dr. Vince Slabe, Hannah Leonard, Dr. Clint Epps, Ev-
erett Headley, Jordan Hoffmaster, Joe Pontecorvo, Chris Parish,
Leland Brown, Russell Kuhlman, Kate Davis, Jake Jourdonnais,
Stephen and Gail Goheen, Bryan Bedrosian, Hannah Nikonow,
and Jeff Gailus.

And thanks to my various shooting and hunting partners who

helped influence my views leading up to this book, including Tanner Banks, Mark Clearwater, Cody Bomberger, Tyler Carlin, and Logan Miller.

Lastly, I'd like to thank my family. To my dad, Greg, thanks for handing me fly rods and weapons before I could recite the alphabet. To my mom, Sue, your support and relentless optimism continually inflates my sails. To my wife, Bridgette, you've endured a decades-worth of me rattling off statistics about lead poisoning. When I holed up in my office or our basement at the tail end of this book project, you remained patient and supportive. And thanks to Norman, who made writing breaks so much fun.

photo © Jordan Hoffmaster

Mike McTee shot his first weapon before he could recite the alphabet. Now, understanding weapons is part of his job.

Mike's career took this trajectory after he gained a B.S. in Environmental Chemistry. Curious about potential pollution at a historic shooting range in Montana's Bitterroot Valley, he earned an M.S. in Geosciences studying the site. Lead contamination soon grabbed Mike's focus. Each winter at MPG Ranch in Western Montana, where Mike works as a wildlife researcher, biologists caught eagles suffering from alarming levels of lead exposure. Mike soon initiated studies on scavenger ecology and began investigating the wound ballistics of rifle bullets, the suspected source of lead. His research has appeared in numerous wildlife and environmental journals.

Mike often connects with the public through his writings and speaking engagements, whether it be to a local group of hunters or a gymnasium full of middle schoolers. He frequently writes about the outdoors, with credits to The FlyFish Journal, Backcountry Journal, Strung, and Bugle. He lives with his family in Missoula, MT.

For further reading about how hunters are defending our natural world, check out our new book: *Teachers in the Forest.*

"The teachings contained within this book belong to the author's decades afield as both hunter and self-provider and are also strongly influenced by his close ties to the Ojibwe people and their connection to the land and its animals." —Traditional Bowhunter Magazine

www.riverfeetpress.com

www.riverfeetpress.com
printed in the USA